# Handbook
# of
# Dosimetry Data
# for
# Radiotherapy

Author

## Shirish K. Jani, Ph.D.

Associate Professor and Chief of Clinical Physics
Division of Radiation Oncology
Department of Radiology
The University of Iowa College of Medicine
Iowa City, Iowa

**CRC Press**
**Boca Raton  Ann Arbor  London  Tokyo**

**Library of Congress Cataloging-in-Publication Data**

Jani, Shirish K.
    Handbook of dosimetry data for radiotherapy / author, Shirish K. Jani.
       p.   cm.
    Includes bibliographical references and index.
    ISBN 0-8493-3263-X
    1. Radiation—Dosimetry.   2. Radiotherapy.   I. Title.
    [DNLM: 1. Radiotherapy Dosage.   2. Radiation Dosage.  WN 450 J33h 1993]
RM849.J35  1993
612'.01448—dc20
DNLM/DLC
for Library of Congress                                   93-1317
                                                    CIP

© 1993 by CRC Press, Inc.

International Standard Book Number 0-8493-3263-X

Library of Congress Card Number 93-1317
Printed in the United States of America   2 3 4 5 6 7 8 9 0
Printed on acid-free paper

# PREFACE

In the last few years, I have felt the need to write a book which would provide typical dosimetric data on several of the radiotherapeutic procedures commonly used today. To my knowledge, no such reference book exists. This book contains typical central axis depth dose data for photon and electron beams, brachytherapy source data, and data on special radiotherapeutic procedures such as total body irradiation (TBI), total skin electron therapy (TSET), stereotactic radiosurgery (SRS), and intraoperative radiotherapy (IORT). Its primary goal is to provide a complete reference for radiation physicists, dosimetrists, and oncologists in their day-to-day routine as well as to aid in teaching clinical dosimetry and physics to trainees in radiation oncology. I believe that this handbook will complement several existing textbooks on radiotherapy physics, and enhance the learning experience.

*This book is dedicated to my parents,
Lalita and Keshavlal Jani, my wife Amita,
and our children Ashish and Shyam*

# THE AUTHOR

**Shirish K. Jani, Ph.D.,** received his B.Sc. in Physics (summa cum laude) from Gujarat University, Ahmedabad, India, in 1973. He earned his M.A. in Physics from The University of Texas at Arlington and his Ph.D. from North Texas State University. In 1980, he joined the Radiation Physics Division, Department of Radiology at the Medical College of Virginia in Richmond, Virginia, first as a postdoctoral fellow, then as a junior staff member. In 1983, he joined the Division of Radiation Oncology at The University of Iowa College of Medicine in Iowa City, Iowa. In 1985, he became chief of the clinical physics section of the Division of Radiation Oncology. For the past twelve years, he has taught clinical physics and dosimetry to radiation oncology residents, technology students, and medical students.

Dr. Jani is certified in therapeutic radiological physics by the American Board of Radiology as well as the American Board of Medical Physics. He has written over 25 scientific papers and two books, and has made numerous presentations on the physics of clinical radiotherapy throughout the nation. He has served as a reviewer for scientific journals such as Medical Physics, Biopolymers, and the American Journal of Bone and Joint Surgery. Dr. Jani is a member of the Physics Commission of the American College of Radiology.

# ACKNOWLEDGMENT

Writing this book was a major project that demanded many months of hard work. I received great support and lots of help from my wonderful colleagues in the Physics section. Kathleen Anderson with all her research skills has helped me from start to finish on this endeavor. Womah Neeranjun prepared most of the illustrations for this book. Judith Wacha, Ed Pennington, and Earl Pelland assisted in editing a large amount of dosimetry data. I am very thankful to all of them for their help. I thank David Hussey, M.D., Director of Radiation Oncology, who made available all the possible resources for this book. At home, my wife, Amita, provided equally important support that was needed to finish this book.

My secretary, Marie Witthoft, has truly been a God-given gift to me. She has helped me tremendously by providing excellent secretarial support. This book would not have been finished without her superb skills and talented approaches. I am very thankful to her for all her help.

# TABLE OF CONTENTS

Chapter 1

# DATA ON PHOTON BEAMS

This chapter contains central axis depth dose data for cobalt-60 beams
and X ray beams in the 3–45-megavolt (MV) energy range. The data include
the dose in the buildup region including surface dose, the percent depth dose
(PDD) at standard source-to-surface distance (SSD), tissue air ratio (TAR) for
Co-60 beams, and the tissue phantom ratio (TPR) for all the photon beams
covering the entire energy range. When the depth of normalization is chosen
to be the depth of dose maximum ($d_{max}$), the tissue phantom ratio is also
called tissue maximum ratio (TMR). All the depth dose data presented have
been measured by users of various treatment units and represent most
commonly used machines in radiotherapy at present. The data are presented
in the following order:

A. Co-60 beams
B. 3–4-MV X rays
C. 6-MV X rays
D. 8–10-MV X rays
E. 14–20-MV X rays
F. 23–25-MV X rays
G. 33–45-MV X rays
H. Surface (skin) dose
I. Dose buildup region
J. Summary of depth dose data

The depth dose values depend primarily on beam energy and field size.
Nominal beam energy in MV is given by the manufacturers. The beam
energy that represents its penetrating power should be determined by
measuring the so-called ionization ratio.[76] The ionization ratio is defined as
the ratio of the ionization measured in water at 20- and 10-cm depths for a
fixed source-to-detector distance. The ionization ratio is then converted to
nominal accelerating potential (NAP). The NAP value for a treatment unit is
normally used in determining parameters for beam calibration.

The depth dose values also depend on the depth at which dose values are
normalized. Ideally, the depth of normalization should be greater than the
largest $d_{max}$ value which occurs at small field sizes.[81] Some data tables
presented in this chapter for a fixed nominal energy have not been normalized
at the same depth. Hence, any comparison of depth doses among various
treatment units must take this into account.

There are three main parameters affecting $d_{max}$, the depth of dose
maximum: energy, distance (SSD) and field size. The $d_{max}$ increases with

beam energy. Likewise, it increases with distance, especially at high beam energies.[80] However, the $d_{max}$ decreases with increasing field size. This shift of $d_{max}$ towards the surface with field size is quite pronounced at high beam energies (>15 MV) and is due either to electron contamination[54] or low-energy photons produced within the accelerator's head.[67]

Relative surface dose (RSD) for photon beams is defined as the dose at or near the surface (i.e., at nearly zero depth) relative to the dose at the depth of dose normalization ($d_n$). For low-energy X rays, the data are usually normalized at $d_{max}$. For high-energy X rays, the $d_{max}$ migrates towards the surface with increasing field size. Therefore, the $d_n$ is usually chosen to be greater than $d_{max}$. This must be considered when comparing surface dose values among different treatment units. The RSD is commonly measured using a parallel plate ionization chamber. If a chamber with fixed plate separation is employed, a depth-dependent correction needs to be applied to ionization readings to determine the dose.[60,73] Significant variation in RSD may exist among X ray beams of same nominal energy but from different treatment units. The reader is referred to some excellent papers on the dose buildup that are listed in Sections H and I of the References for this chapter.

The surface dose for photon beams is usually quite low and therefore a large skin-sparing effect is observed for most photon treatments. Placement of any plastic tray or wedge filter may alter the surface dose significantly.[56] Moreover, the dose buildup is very rapid for low- and medium-energy megavoltage beams, especially at large field sizes.

Wedge filters are occasionally used to systematically alter the dose distribution within a megavoltage beam so that a combination of two or more beams aimed at the tumor will result in a homogeneous dose throughout the treatment volume. The wedge attenuation factors depend somewhat upon the field size and the depth of measurement.[82] In general, wedge filters do not alter the depth dose characteristics of the beam. Therefore, the data for wedged fields are not presented here.

It is well known that standard methods of computing the equivalent square of a rectangular field do not always yield correct answers. For example, Sterling's formula[83] may introduce errors in depth dose determination when used for highly elongated fields.[77,80] It is standard practice to determine equivalence of square and rectangular fields by making depth dose measurements at the time of commissioning a treatment machine. Such equivalence tables are not included in this book.

A summary of depth dose data for X and gamma ray beams is presented in Table 1.J1. The reference section at the end of the chapter lists articles separately for each energy range.

## LIST OF TABLES

## TABLE 1.A1
### Percent Depth Doses for Cobalt-60 Gamma Rays [AECL Theratron-80, 80 cm SSD]

| Depth (cm) | Field Size (cm×cm) | | | | | | | | | | |
|---|---|---|---|---|---|---|---|---|---|---|---|
| | 4×4 | 5×5 | 6×6 | 8×8 | 10×10 | 12×12 | 15×15 | 20×20 | 25×25 | 30×30 | 32×32 |
| 0.5 | 100.0 | 100.0 | 100.0 | 100.0 | 100.0 | 100.0 | 100.0 | 100.0 | 100.0 | 100.0 | 100.0 |
| 1.0 | 97.2 | 97.5 | 97.7 | 97.9 | 98.1 | 98.2 | 98.3 | 98.3 | 98.4 | 98.5 | 98.5 |
| 2.0 | 91.4 | 92.1 | 93.2 | 93.6 | 93.6 | 93.9 | 94.1 | 94.3 | 94.5 | 94.7 | 94.7 |
| 3.0 | 85.4 | 86.3 | 88.0 | 88.6 | 88.6 | 89.1 | 89.5 | 90.0 | 90.3 | 90.5 | 90.5 |
| 4.0 | 79.7 | 80.7 | 82.8 | 83.6 | 83.6 | 84.2 | 84.8 | 85.5 | 86.0 | 86.3 | 86.4 |
| 5.0 | 73.9 | 75.1 | 77.7 | 78.7 | 78.7 | 79.4 | 80.2 | 81.2 | 81.7 | 82.0 | 82.2 |
| 6.0 | 68.4 | 69.7 | 72.6 | 73.8 | 73.8 | 74.8 | 75.8 | 76.8 | 77.5 | 78.0 | 78.1 |
| 7.0 | 63.4 | 64.7 | 67.7 | 69.1 | 69.1 | 70.1 | 71.3 | 72.5 | 73.3 | 73.9 | 74.0 |
| 8.0 | 58.5 | 59.9 | 63.1 | 64.6 | 64.6 | 65.7 | 67.0 | 68.4 | 69.4 | 70.0 | 70.2 |
| 9.0 | 54.0 | 55.4 | 58.8 | 60.4 | 60.4 | 61.6 | 62.9 | 64.5 | 65.5 | 66.2 | 66.4 |
| 10.0 | 49.7 | 51.2 | 54.7 | 56.3 | 56.3 | 57.6 | 59.0 | 60.7 | 61.8 | 62.5 | 62.7 |
| 12.0 | 42.4 | 43.8 | 47.1 | 48.8 | 48.8 | 50.2 | 51.7 | 53.6 | 54.8 | 55.7 | 55.9 |
| 14.0 | 36.1 | 37.4 | 40.6 | 42.3 | 42.3 | 43.6 | 45.2 | 47.2 | 48.6 | 49.5 | 49.8 |
| 16.0 | 30.8 | 31.9 | 35.0 | 36.7 | 36.7 | 38.0 | 39.6 | 41.7 | 43.1 | 44.0 | 44.3 |
| 18.0 | 26.4 | 27.4 | 30.2 | 31.8 | 31.8 | 33.1 | 34.7 | 36.7 | 38.1 | 39.0 | 39.3 |
| 20.0 | 22.2 | 23.2 | 25.8 | 27.3 | 27.3 | 28.6 | 30.1 | 32.1 | 33.4 | 34.4 | 34.7 |
| 22.0 | 19.1 | 20.0 | 22.4 | 23.8 | 23.8 | 25.0 | 26.4 | 28.3 | 29.6 | 30.6 | 30.9 |
| 24.0 | 16.2 | 17.0 | 19.2 | 20.5 | 20.5 | 21.6 | 23.0 | 24.8 | 26.1 | 27.0 | 27.4 |
| 26.0 | 14.0 | 14.7 | 16.7 | 17.9 | 17.9 | 19.0 | 20.3 | 22.0 | 23.2 | 24.1 | 24.4 |
| 28.0 | 12.0 | 12.7 | 14.5 | 15.5 | 15.5 | 16.5 | 17.7 | 19.4 | 20.5 | 21.3 | 21.6 |
| 30.0 | 10.1 | 10.7 | 12.3 | 13.3 | 13.3 | 14.2 | 15.3 | 16.9 | 18.0 | 18.7 | 19.0 |

Adapted from The University of Iowa Hospitals and Clinics, Iowa City, IA, 1992.

**TABLE 1.A2**
**Tissue Air Ratios for Cobalt-60 Gamma Rays [AECL Theratron-80, 80 cm SAD]**

| Depth (cm) | Field Size (cm×cm) | | | | | | | | | |
|---|---|---|---|---|---|---|---|---|---|---|
| | 4×4 | 5×5 | 6×6 | 8×8 | 10×10 | 12×12 | 15×15 | 20×20 | 24×24 | 30×30 |
| 0.5 | 1.015 | 1.018 | 1.022 | 1.029 | 1.035 | 1.041 | 1.051 | 1.063 | 1.071 | 1.080 |
| 1.0 | 0.996 | 1.003 | 1.009 | 1.021 | 1.031 | 1.038 | 1.048 | 1.062 | 1.070 | 1.079 |
| 2.0 | 0.956 | 0.967 | 0.976 | 0.992 | 1.004 | 1.014 | 1.025 | 1.040 | 1.049 | 1.059 |
| 3.0 | 0.915 | 0.928 | 0.940 | 0.959 | 0.974 | 0.985 | 0.999 | 1.016 | 1.027 | 1.038 |
| 4.0 | 0.872 | 0.888 | 0.902 | 0.924 | 0.940 | 0.953 | 0.968 | 0.987 | 1.000 | 1.014 |
| 5.0 | 0.829 | 0.847 | 0.862 | 0.887 | 0.905 | 0.919 | 0.936 | 0.957 | 0.972 | 0.988 |
| 6.0 | 0.786 | 0.805 | 0.821 | 0.847 | 0.867 | 0.883 | 0.902 | 0.925 | 0.942 | 0.959 |
| 7.0 | 0.743 | 0.762 | 0.778 | 0.807 | 0.827 | 0.845 | 0.866 | 0.893 | 0.911 | 0.929 |
| 8.0 | 0.700 | 0.719 | 0.736 | 0.765 | 0.787 | 0.806 | 0.830 | 0.859 | 0.879 | 0.899 |
| 9.0 | 0.659 | 0.677 | 0.695 | 0.724 | 0.747 | 0.768 | 0.793 | 0.825 | 0.847 | 0.869 |
| 10.0 | 0.620 | 0.638 | 0.665 | 0.685 | 0.709 | 0.730 | 0.756 | 0.790 | 0.813 | 0.837 |
| 12.0 | 0.546 | 0.563 | 0.580 | 0.611 | 0.636 | 0.658 | 0.685 | 0.722 | 0.747 | 0.772 |
| 14.0 | 0.482 | 0.499 | 0.515 | 0.545 | 0.571 | 0.594 | 0.622 | 0.660 | 0.701 | 0.715 |
| 16.0 | 0.427 | 0.443 | 0.458 | 0.485 | 0.510 | 0.533 | 0.564 | 0.605 | 0.632 | 0.660 |
| 18.0 | 0.378 | 0.393 | 0.406 | 0.433 | 0.457 | 0.479 | 0.509 | 0.551 | 0.579 | 0.607 |
| 20.0 | 0.333 | 0.347 | 0.361 | 0.386 | 0.410 | 0.431 | 0.461 | 0.502 | 0.531 | 0.560 |

Adapted from The University of Iowa Hospitals and Clinics, Iowa City, IA, 1992.

## TABLE 1.A3
## Tissue Phantom Ratios for Cobalt-60 Gamma Rays [AECL Theratron-80, 80 cm SAD]

| Depth (cm) | Field Size (cm×cm) | | | | | | | | | | |
|---|---|---|---|---|---|---|---|---|---|---|---|
| | 4×4 | 5×5 | 6×6 | 8×8 | 10×10 | 12×12 | 15×15 | 20×20 | 25×25 | 30×30 | 32×32 |
| 0.5 | 1.000 | 1.000 | 1.000 | 1.000 | 1.000 | 1.000 | 1.000 | 1.000 | 1.000 | 1.000 | 1.000 |
| 1.0 | 0.984 | 0.987 | 0.989 | 0.991 | 0.993 | 0.994 | 0.995 | 0.995 | 0.996 | 0.997 | 0.997 |
| 2.0 | 0.947 | 0.954 | 0.959 | 0.966 | 0.970 | 0.973 | 0.975 | 0.978 | 0.980 | 0.982 | 0.982 |
| 3.0 | 0.906 | 0.915 | 0.922 | 0.933 | 0.940 | 0.945 | 0.950 | 0.955 | 0.959 | 0.961 | 0.961 |
| 4.0 | 0.865 | 0.875 | 0.885 | 0.898 | 0.907 | 0.914 | 0.921 | 0.929 | 0.934 | 0.937 | 0.939 |
| 5.0 | 0.820 | 0.833 | 0.844 | 0.862 | 0.873 | 0.881 | 0.890 | 0.901 | 0.908 | 0.912 | 0.914 |
| 6.0 | 0.776 | 0.789 | 0.802 | 0.822 | 0.836 | 0.847 | 0.859 | 0.872 | 0.880 | 0.886 | 0.888 |
| 7.0 | 0.734 | 0.748 | 0.762 | 0.783 | 0.799 | 0.812 | 0.825 | 0.841 | 0.851 | 0.859 | 0.861 |
| 8.0 | 0.692 | 0.707 | 0.721 | 0.745 | 0.762 | 0.776 | 0.791 | 0.810 | 0.822 | 0.831 | 0.834 |
| 9.0 | 0.652 | 0.668 | 0.683 | 0.708 | 0.727 | 0.742 | 0.758 | 0.779 | 0.793 | 0.802 | 0.805 |
| 10.0 | 0.612 | 0.628 | 0.644 | 0.671 | 0.691 | 0.707 | 0.726 | 0.747 | 0.763 | 0.773 | 0.777 |
| 12.0 | 0.543 | 0.559 | 0.574 | 0.600 | 0.621 | 0.639 | 0.660 | 0.685 | 0.703 | 0.717 | 0.721 |
| 14.0 | 0.481 | 0.495 | 0.510 | 0.537 | 0.558 | 0.576 | 0.598 | 0.626 | 0.647 | 0.662 | 0.667 |
| 16.0 | 0.427 | 0.439 | 0.452 | 0.477 | 0.499 | 0.519 | 0.542 | 0.571 | 0.593 | 0.610 | 0.615 |
| 18.0 | 0.378 | 0.390 | 0.403 | 0.426 | 0.448 | 0.467 | 0.490 | 0.520 | 0.543 | 0.560 | 0.565 |
| 20.0 | 0.329 | 0.341 | 0.353 | 0.376 | 0.396 | 0.415 | 0.438 | 0.468 | 0.492 | 0.509 | 0.515 |
| 22.0 | 0.293 | 0.304 | 0.315 | 0.336 | 0.355 | 0.373 | 0.396 | 0.426 | 0.449 | 0.467 | 0.473 |
| 24.0 | 0.257 | 0.267 | 0.277 | 0.296 | 0.314 | 0.331 | 0.354 | 0.384 | 0.407 | 0.425 | 0.431 |
| 26.0 | 0.230 | 0.239 | 0.249 | 0.266 | 0.283 | 0.299 | 0.320 | 0.349 | 0.372 | 0.390 | 0.396 |
| 28.0 | 0.203 | 0.212 | 0.220 | 0.236 | 0.252 | 0.267 | 0.287 | 0.315 | 0.338 | 0.356 | 0.362 |
| 30.0 | 0.176 | 0.184 | 0.192 | 0.206 | 0.221 | 0.234 | 0.253 | 0.281 | 0.303 | 0.321 | 0.327 |

Adapted from The University of Iowa Hospitals and Clinics, Iowa City, IA, 1992.

## TABLE 1.A4
## Percent Depth Doses for Cobalt-60 Gamma Rays [Theratronics Theratron-1000, 100 cm SSD]

| Depth (cm) | Field Size (cm×cm) | | | | | | | | | | | |
|---|---|---|---|---|---|---|---|---|---|---|---|---|
| | 6×6 | 8×8 | 10×10 | 12×12 | 15×15 | 18×18 | 20×20 | 25×25 | 30×30 | 40×40 |
| 0.4 | 100.0 | 100.0 | 100.0 | 100.0 | 100.0 | 100.0 | 100.0 | 100.0 | 100.0 | 100.0 |
| 1.0 | 98.1 | 98.6 | 98.2 | 98.2 | 98.4 | 98.3 | 98.5 | 99.3 | 99.0 | 98.4 |
| 2.0 | 93.6 | 94.1 | 94.2 | 94.2 | 94.6 | 94.6 | 94.8 | 95.8 | 95.7 | 95.2 |
| 3.0 | 88.5 | 89.4 | 89.5 | 90.0 | 90.5 | 90.8 | 90.9 | 92.2 | 92.1 | 91.5 |
| 4.0 | 83.1 | 84.3 | 85.0 | 85.6 | 86.3 | 86.5 | 86.9 | 88.2 | 88.3 | 88.1 |
| 5.0 | 77.9 | 79.5 | 80.3 | 80.9 | 81.8 | 82.5 | 82.7 | 84.3 | 84.3 | 84.2 |
| 6.0 | 72.7 | 74.4 | 75.8 | 76.4 | 77.7 | 78.2 | 78.7 | 80.2 | 80.2 | 80.4 |
| 7.0 | 67.6 | 69.9 | 71.2 | 72.1 | 73.5 | 74.3 | 74.9 | 76.3 | 76.4 | 76.8 |
| 8.0 | 62.8 | 65.3 | 66.7 | 67.8 | 69.4 | 70.2 | 70.7 | 72.5 | 72.7 | 73.1 |
| 9.0 | 58.5 | 61.1 | 62.7 | 63.9 | 65.3 | 66.5 | 67.0 | 68.8 | 69.0 | 69.7 |
| 10.0 | 54.3 | 56.9 | 58.7 | 59.7 | 61.6 | 62.7 | 63.3 | 65.0 | 65.5 | 66.2 |
| 12.0 | 46.8 | 49.2 | 51.2 | 52.5 | 54.4 | 55.7 | 56.4 | 58.3 | 58.7 | 59.4 |
| 14.0 | 40.0 | 42.6 | 44.6 | 45.9 | 47.9 | 49.3 | 50.2 | 51.8 | 52.6 | 53.5 |
| 16.0 | 34.3 | 36.8 | 38.5 | 40.0 | 42.0 | 43.6 | 44.3 | 45.9 | 46.6 | 47.8 |
| 18.0 | 29.5 | 31.7 | 33.4 | 34.7 | 36.6 | 38.3 | 39.1 | 40.7 | 41.5 | 42.6 |
| 20.0 | 25.3 | 27.1 | 28.9 | 30.1 | 32.1 | 33.5 | 34.3 | 36.0 | 36.8 | 37.8 |
| 22.0 | 21.6 | 23.4 | 25.0 | 26.3 | 28.1 | 29.5 | 30.2 | 31.7 | 32.6 | 33.8 |
| 24.0 | 18.6 | 20.0 | 21.6 | 22.7 | 24.4 | 25.7 | 26.6 | 28.0 | 28.7 | 30.2 |
| 26.0 | 15.9 | 17.2 | 18.5 | 19.4 | 21.2 | 22.5 | 23.2 | 24.6 | 25.5 | 26.7 |
| 28.0 | 13.4 | 14.9 | 15.9 | 16.7 | 18.4 | 19.5 | 20.3 | 21.6 | 22.3 | 23.5 |
| 30.0 | 11.6 | 12.7 | 13.6 | 14.5 | 15.9 | 17.3 | 17.8 | 18.9 | 19.7 | 20.9 |

From Glasgow, G., Loyola University, Chicago, Illinois, personal communication, 1992. With permission.

TABLE 1.A5
Tissue Air Ratios for Cobalt-60 Gamma Rays [Theratronics Theratron-1000, 100 cm SAD]

| Depth (cm) | Field Size (cm×cm) | | | | | | | | | |
|---|---|---|---|---|---|---|---|---|---|---|
| | 6×6 | 8×8 | 10×10 | 12×12 | 15×15 | 18×18 | 20×20 | 25×25 | 30×30 | 40×40 |
| 0.4 | 1.024 | 1.030 | 1.036 | 1.043 | 1.049 | 1.055 | 1.058 | 1.064 | 1.069 | 1.075 |
| 1.0 | 1.016 | 1.028 | 1.030 | 1.037 | 1.045 | 1.050 | 1.054 | 1.070 | 1.071 | 1.070 |
| 2.0 | 0.989 | 1.000 | 1.007 | 1.014 | 1.024 | 1.030 | 1.035 | 1.051 | 1.056 | 1.056 |
| 3.0 | 0.953 | 0.968 | 0.976 | 0.986 | 0.999 | 1.006 | 1.012 | 1.031 | 1.036 | 1.036 |
| 4.0 | 0.912 | 0.930 | 0.942 | 0.956 | 0.969 | 0.978 | 0.984 | 1.006 | 1.012 | 1.014 |
| 5.0 | 0.870 | 0.891 | 0.907 | 0.919 | 0.936 | 0.949 | 0.955 | 0.977 | 0.985 | 0.988 |
| 6.0 | 0.826 | 0.849 | 0.870 | 0.884 | 0.904 | 0.917 | 0.924 | 0.948 | 0.956 | 0.960 |
| 7.0 | 0.780 | 0.810 | 0.832 | 0.848 | 0.870 | 0.885 | 0.894 | 0.918 | 0.924 | 0.934 |
| 8.0 | 0.737 | 0.769 | 0.792 | 0.810 | 0.836 | 0.851 | 0.859 | 0.884 | 0.897 | 0.903 |
| 9.0 | 0.698 | 0.731 | 0.756 | 0.776 | 0.798 | 0.818 | 0.829 | 0.853 | 0.867 | 0.873 |
| 10.0 | 0.658 | 0.690 | 0.718 | 0.738 | 0.766 | 0.784 | 0.795 | 0.820 | 0.835 | 0.846 |
| 12.0 | 0.586 | 0.616 | 0.644 | 0.668 | 0.697 | 0.718 | 0.730 | 0.758 | 0.776 | 0.786 |
| 14.0 | 0.514 | 0.548 | 0.578 | 0.601 | 0.631 | 0.654 | 0.668 | 0.695 | 0.715 | 0.729 |
| 16.0 | 0.454 | 0.486 | 0.514 | 0.537 | 0.566 | 0.593 | 0.608 | 0.634 | 0.655 | 0.674 |
| 18.0 | 0.402 | 0.431 | 0.458 | 0.480 | 0.508 | 0.532 | 0.547 | 0.578 | 0.601 | 0.620 |
| 20.0 | 0.356 | 0.380 | 0.404 | 0.428 | 0.455 | 0.481 | 0.494 | 0.523 | 0.548 | 0.568 |
| 22.0 | 0.311 | 0.335 | 0.359 | 0.380 | 0.408 | 0.432 | 0.447 | 0.474 | 0.496 | 0.520 |
| 24.0 | 0.277 | 0.296 | 0.315 | 0.337 | 0.362 | 0.386 | 0.399 | 0.430 | 0.452 | 0.475 |
| 26.0 | 0.243 | 0.261 | 0.278 | 0.296 | 0.318 | 0.345 | 0.355 | 0.385 | 0.406 | 0.431 |
| 28.0 | 0.206 | 0.226 | 0.247 | 0.262 | 0.281 | 0.307 | 0.319 | 0.346 | 0.367 | 0.389 |
| 30.0 | 0.185 | 0.201 | 0.216 | 0.230 | 0.250 | 0.271 | 0.283 | 0.312 | 0.329 | 0.353 |

From Glasgow, G., Loyola University, Chicago, Illinois, personal communication, 1992. With permission.

## TABLE 1.B1
### Percent Depth Doses for 3.3-MV X Rays [Siemens 5800, 100 cm SSD]

| Depth (cm) | Field Size (cm×cm) | | | | | | | |
|---|---|---|---|---|---|---|---|---|
| | 4×4 | 6×6 | 8×8 | 10×10 | 15×15 | 20×20 | 25×25 | 40×40 |
| 0.7 | 100.0 | 100.0 | 100.0 | 100.0 | 100.0 | 100.0 | 100.0 | 100.0 |
| 1.0 | 100.1 | 100.1 | 100.0 | 100.2 | 100.0 | 100.0 | 99.9 | 99.7 |
| 2.0 | 94.6 | 95.1 | 95.5 | 96.1 | 96.4 | 96.7 | 96.8 | 96.9 |
| 3.0 | 88.5 | 89.8 | 90.7 | 91.3 | 92.4 | 93.2 | 93.7 | 93.8 |
| 4.0 | 82.1 | 84.1 | 85.5 | 86.6 | 88.1 | 88.9 | 89.8 | 90.1 |
| 5.0 | 76.1 | 78.7 | 80.4 | 81.8 | 83.6 | 85.0 | 85.9 | 86.6 |
| 6.0 | 70.4 | 73.1 | 75.4 | 77.0 | 79.4 | 81.3 | 82.1 | 83.0 |
| 7.0 | 64.9 | 67.8 | 70.3 | 72.1 | 74.9 | 76.9 | 78.1 | 79.1 |
| 8.0 | 59.8 | 63.0 | 65.5 | 67.7 | 70.7 | 72.9 | 74.4 | 75.4 |
| 9.0 | 55.4 | 58.4 | 61.0 | 63.2 | 66.7 | 69.2 | 70.5 | 72.0 |
| 10.0 | 50.9 | 54.1 | 56.9 | 58.9 | 62.7 | 65.4 | 66.9 | 68.3 |
| 11.0 | 47.0 | 50.1 | 52.7 | 55.0 | 58.9 | 61.6 | 63.4 | 65.0 |
| 12.0 | 43.4 | 46.2 | 49.0 | 51.3 | 55.5 | 58.1 | 59.9 | 61.7 |
| 13.0 | 39.8 | 42.8 | 45.6 | 47.9 | 51.7 | 54.8 | 56.6 | 58.6 |
| 14.0 | 36.8 | 39.5 | 42.1 | 44.5 | 48.5 | 51.5 | 53.5 | 55.4 |
| 15.0 | 34.0 | 36.6 | 39.3 | 41.4 | 45.4 | 48.5 | 50.4 | 52.6 |
| 16.0 | 31.3 | 33.9 | 36.3 | 38.5 | 42.5 | 45.4 | 47.5 | 49.6 |
| 17.0 | 28.9 | 31.4 | 33.8 | 35.7 | 39.8 | 42.7 | 44.6 | 46.9 |
| 18.0 | 26.7 | 29.0 | 31.4 | 33.3 | 37.1 | 40.2 | 42.0 | 44.3 |
| 19.0 | 24.7 | 26.8 | 29.1 | 30.9 | 34.9 | 37.9 | 39.7 | 41.8 |
| 20.0 | 22.7 | 24.7 | 27.0 | 28.9 | 32.5 | 35.3 | 37.2 | 39.5 |
| 22.0 | 19.3 | 21.2 | 23.1 | 24.9 | 28.3 | 31.0 | 33.0 | 35.2 |
| 24.0 | 16.5 | 18.1 | 19.9 | 21.5 | 24.7 | 27.2 | 29.0 | 31.3 |
| 26.0 | 14.1 | 15.4 | 17.0 | 18.5 | 21.6 | 23.8 | 25.6 | 27.8 |
| 28.0 | 12.0 | 13.2 | 14.7 | 15.9 | 18.6 | 20.8 | 22.4 | 24.7 |
| 30.0 | 10.4 | 11.3 | 12.7 | 13.8 | 16.2 | 18.4 | 19.8 | 21.7 |

From Steidley, K.D. and Rosen C.W., Dosimetric aspects of a 3.3-MV linear accelerator, *Med. Phys.*, 17, 474, 1990. With permission.

**TABLE 1.B2**

**Percent Depth Doses for 4-MV X Rays [Varian Clinac-4/80, 80 cm SSD]**

| Depth (cm) | Field Size (cm×cm) | | | | | | | | | | |
|---|---|---|---|---|---|---|---|---|---|---|---|
| | 4×4 | 5×5 | 6×6 | 8×8 | 10×10 | 12×12 | 15×15 | 20×20 | 25×25 | 30×30 | 32×32 |
| 1.0 | 100.0 | 100.0 | 100.0 | 100.0 | 100.0 | 100.0 | 100.0 | 100.0 | 100.0 | 100.0 | 100.0 |
| 2.0 | 94.9 | 95.2 | 95.4 | 95.6 | 95.7 | 95.8 | 96.0 | 96.4 | 96.3 | 96.2 | 96.2 |
| 3.0 | 88.9 | 89.5 | 89.9 | 90.6 | 91.0 | 91.3 | 91.5 | 91.9 | 92.4 | 92.5 | 92.5 |
| 4.0 | 83.1 | 83.9 | 84.4 | 85.4 | 86.0 | 86.5 | 87.0 | 87.6 | 88.3 | 88.4 | 88.4 |
| 5.0 | 77.5 | 78.5 | 79.2 | 80.4 | 81.2 | 81.9 | 82.6 | 83.6 | 84.4 | 84.5 | 84.4 |
| 6.0 | 72.2 | 73.3 | 74.2 | 75.6 | 76.7 | 77.5 | 78.4 | 79.6 | 80.6 | 81.0 | 81.0 |
| 7.0 | 67.1 | 68.4 | 69.4 | 71.1 | 72.4 | 73.3 | 74.4 | 75.8 | 77.0 | 77.6 | 77.8 |
| 8.0 | 62.5 | 63.7 | 64.9 | 66.7 | 68.1 | 69.1 | 70.3 | 71.9 | 73.2 | 73.9 | 74.0 |
| 9.0 | 58.0 | 59.3 | 60.5 | 62.4 | 64.0 | 65.1 | 66.4 | 68.2 | 69.6 | 70.3 | 70.4 |
| 10.0 | 53.7 | 55.1 | 56.3 | 58.4 | 60.0 | 61.3 | 62.7 | 64.6 | 66.1 | 66.8 | 66.9 |
| 12.0 | 46.2 | 47.5 | 48.8 | 50.9 | 52.6 | 54.0 | 55.7 | 58.1 | 59.9 | 60.7 | 60.9 |
| 14.0 | 40.0 | 41.2 | 42.4 | 44.4 | 46.2 | 47.7 | 49.5 | 51.9 | 53.7 | 54.6 | 54.8 |
| 16.0 | 34.5 | 35.7 | 36.8 | 38.8 | 40.5 | 42.0 | 43.8 | 46.3 | 48.1 | 49.0 | 49.3 |
| 18.0 | 29.6 | 30.8 | 31.9 | 33.7 | 35.4 | 36.9 | 38.7 | 41.2 | 43.1 | 43.9 | 44.2 |
| 20.0 | 25.5 | 26.6 | 27.6 | 29.3 | 30.9 | 32.5 | 34.2 | 36.5 | 38.3 | 39.3 | 39.6 |
| 22.0 | 22.1 | 23.1 | 24.1 | 25.6 | 27.1 | 28.6 | 30.3 | 32.6 | 34.2 | 35.0 | 35.4 |
| 24.0 | 19.1 | 20.1 | 20.9 | 22.3 | 23.7 | 25.1 | 26.7 | 28.9 | 30.4 | 31.3 | 31.7 |
| 26.0 | 16.7 | 17.6 | 18.4 | 19.6 | 20.9 | 22.1 | 23.6 | 25.7 | 27.2 | 28.1 | 28.5 |
| 28.0 | 14.5 | 15.4 | 16.1 | 17.1 | 18.4 | 19.5 | 20.8 | 22.9 | 24.3 | 25.2 | 25.6 |
| 30.0 | 12.5 | 13.4 | 13.9 | 14.9 | 16.1 | 16.9 | 18.1 | 20.3 | 21.6 | 22.5 | 22.9 |

Adapted from The University of Iowa Hospitals and Clinics, Iowa City, IA, 1992, personal communication.

*Handbook of Dosimetry Data for Radiotherapy*

**TABLE 1.B3**

**Tissue Phantom Ratios for 4-MV X Rays [Varian Clinac-4/80, 80 cm SAD]**

| Depth (cm) | Field Size (cm×cm) | | | | | | | | | | | |
|---|---|---|---|---|---|---|---|---|---|---|---|---|
| | 4×4 | 5×5 | 6×6 | 8×8 | 10×10 | 12×12 | 15×15 | 20×20 | 25×25 | 30×30 | 32×32 |
| 1.0 | 1.000 | 1.000 | 1.000 | 1.000 | 1.000 | 1.000 | 1.000 | 1.000 | 1.000 | 1.000 | 1.000 |
| 2.0 | 0.971 | 0.975 | 0.977 | 0.980 | 0.980 | 0.981 | 0.983 | 0.987 | 0.987 | 0.986 | 0.986 |
| 3.0 | 0.932 | 0.938 | 0.943 | 0.950 | 0.954 | 0.957 | 0.960 | 0.963 | 0.969 | 0.970 | 0.971 |
| 4.0 | 0.891 | 0.899 | 0.906 | 0.916 | 0.922 | 0.928 | 0.933 | 0.940 | 0.948 | 0.950 | 0.950 |
| 5.0 | 0.849 | 0.860 | 0.868 | 0.882 | 0.891 | 0.898 | 0.906 | 0.917 | 0.926 | 0.929 | 0.929 |
| 6.0 | 0.808 | 0.820 | 0.831 | 0.848 | 0.859 | 0.868 | 0.879 | 0.892 | 0.904 | 0.910 | 0.911 |
| 7.0 | 0.767 | 0.781 | 0.794 | 0.814 | 0.828 | 0.839 | 0.852 | 0.868 | 0.881 | 0.891 | 0.893 |
| 8.0 | 0.729 | 0.743 | 0.757 | 0.779 | 0.794 | 0.807 | 0.822 | 0.840 | 0.855 | 0.866 | 0.869 |
| 9.0 | 0.691 | 0.706 | 0.720 | 0.744 | 0.761 | 0.776 | 0.792 | 0.812 | 0.830 | 0.842 | 0.842 |
| 10.0 | 0.653 | 0.668 | 0.683 | 0.709 | 0.728 | 0.744 | 0.762 | 0.785 | 0.804 | 0.818 | 0.821 |
| 12.0 | 0.584 | 0.599 | 0.615 | 0.642 | 0.662 | 0.680 | 0.702 | 0.731 | 0.755 | 0.773 | 0.778 |
| 14.0 | 0.524 | 0.540 | 0.555 | 0.581 | 0.602 | 0.621 | 0.646 | 0.677 | 0.703 | 0.722 | 0.728 |
| 16.0 | 0.469 | 0.484 | 0.499 | 0.525 | 0.546 | 0.565 | 0.591 | 0.624 | 0.652 | 0.673 | 0.679 |
| 18.0 | 0.419 | 0.432 | 0.446 | 0.473 | 0.493 | 0.512 | 0.539 | 0.573 | 0.602 | 0.625 | 0.631 |
| 20.0 | 0.374 | 0.386 | 0.399 | 0.425 | 0.444 | 0.464 | 0.491 | 0.524 | 0.553 | 0.576 | 0.583 |
| 22.0 | 0.336 | 0.347 | 0.359 | 0.383 | 0.401 | 0.420 | 0.447 | 0.481 | 0.510 | 0.532 | 0.539 |
| 24.0 | 0.299 | 0.310 | 0.323 | 0.345 | 0.362 | 0.379 | 0.405 | 0.439 | 0.467 | 0.490 | 0.497 |
| 26.0 | 0.264 | 0.278 | 0.292 | 0.313 | 0.328 | 0.344 | 0.369 | 0.400 | 0.428 | 0.451 | 0.458 |
| 28.0 | 0.230 | 0.248 | 0.263 | 0.284 | 0.296 | 0.312 | 0.335 | 0.364 | 0.393 | 0.416 | 0.423 |
| 30.0 | 0.196 | 0.218 | 0.235 | 0.254 | 0.265 | 0.280 | 0.301 | 0.327 | 0.358 | 0.381 | 0.388 |

Adapted from The University of Iowa Hospitals and Clinics, Iowa City, IA, 1992, personal communication.

**TABLE 1.B4**

**Percent Depth Doses for 4-MV X Rays [Varian Clinac-4/100, 100 cm SSD]**

| Depth (cm) | Field Size (cm×cm) | | | | | | | | | | | | | | | |
|---|---|---|---|---|---|---|---|---|---|---|---|---|---|---|---|---|
| | 4×4 | 5×5 | 6×6 | 8×8 | 10×10 | 12×12 | 15×15 | 18×18 | 20×20 | 22×22 | 25×25 | 30×30 | 32×32 | 35×35 | 38×38 | 40×40 |
| 1 | 100.0 | 100.0 | 100.0 | 100.0 | 100.0 | 100.0 | 100.0 | 100.0 | 100.0 | 100.0 | 100.0 | 100.0 | 100.0 | 100.0 | 100.0 | 100.0 |
| 2 | 97.3 | 97.3 | 97.4 | 97.5 | 97.6 | 97.6 | 97.7 | 97.8 | 97.8 | 97.8 | 97.9 | 97.9 | 97.9 | 98.0 | 98.1 | 99.1 |
| 3 | 92.0 | 92.7 | 93.2 | 93.8 | 94.1 | 94.3 | 94.4 | 94.5 | 94.5 | 94.6 | 94.7 | 94.7 | 94.7 | 94.8 | 94.8 | 94.9 |
| 4 | 86.8 | 87.5 | 88.1 | 89.0 | 89.5 | 89.9 | 90.2 | 90.4 | 90.6 | 90.7 | 90.9 | 91.1 | 91.1 | 91.2 | 91.3 | 91.3 |
| 5 | 81.8 | 82.3 | 82.9 | 84.0 | 84.8 | 85.5 | 86.1 | 86.6 | 86.9 | 87.0 | 87.2 | 87.5 | 87.6 | 87.7 | 87.8 | 87.8 |
| 6 | 76.6 | 77.3 | 78.1 | 79.4 | 80.5 | 81.3 | 82.1 | 82.7 | 83.0 | 83.2 | 83.5 | 83.9 | 84.1 | 84.2 | 84.4 | 84.5 |
| 8 | 67.1 | 68.2 | 69.1 | 70.6 | 71.8 | 72.8 | 73.9 | 74.8 | 75.2 | 75.6 | 76.1 | 76.7 | 77.0 | 77.3 | 77.6 | 77.8 |
| 10 | 58.5 | 59.1 | 60.9 | 62.8 | 64.2 | 65.3 | 66.5 | 67.5 | 68.0 | 68.4 | 69.0 | 69.8 | 70.0 | 70.4 | 70.8 | 71.1 |
| 12 | 51.6 | 52.8 | 53.8 | 55.5 | 56.9 | 58.2 | 59.7 | 60.8 | 61.4 | 61.9 | 62.6 | 63.5 | 63.8 | 64.2 | 64.6 | 64.9 |
| 14 | 44.7 | 45.8 | 46.9 | 48.7 | 50.3 | 51.6 | 53.1 | 54.4 | 55.1 | 55.7 | 56.5 | 57.5 | 57.9 | 58.4 | 59.0 | 59.4 |
| 17 | 36.8 | 37.9 | 38.8 | 40.4 | 41.9 | 43.2 | 44.8 | 46.1 | 46.8 | 47.4 | 48.2 | 49.3 | 49.7 | 50.3 | 51.0 | 51.4 |
| 20 | 29.9 | 30.9 | 31.9 | 33.5 | 35.0 | 36.2 | 37.8 | 39.1 | 39.8 | 40.4 | 41.1 | 42.2 | 42.7 | 43.3 | 43.9 | 44.4 |
| 24 | 22.7 | 23.7 | 24.5 | 26.0 | 27.2 | 28.4 | 30.0 | 31.2 | 32.0 | 32.6 | 33.4 | 34.5 | 34.9 | 35.5 | 36.2 | 36.6 |
| 28 | 17.4 | 18.2 | 18.9 | 20.2 | 21.3 | 22.4 | 23.8 | 24.8 | 25.4 | 25.9 | 26.6 | 27.6 | 27.9 | 28.5 | 29.1 | 29.5 |
| 32 | 13.4 | 14.0 | 14.7 | 15.8 | 16.8 | 17.8 | 19.0 | 20.0 | 20.5 | 20.9 | 21.5 | 22.4 | 22.7 | 23.3 | 23.8 | 24.1 |
| 36 | 10.3 | 10.8 | 11.2 | 12.1 | 13.0 | 13.9 | 15.0 | 15.9 | 16.3 | 16.7 | 17.3 | 18.0 | 18.3 | 18.8 | 19.2 | 19.5 |

From Biggs, P.J., Doppke, K.P., Leong, J.C., and Russell, M.D., Tissue phantom ratios for a Clinac 4/100, *Med. Phys.*, 9, 753, 1982. With permission.

## TABLE 1.B5
### Tissue Phantom Ratios for 4-MV X Rays [Varian Clinac-4/100, 100 cm SAD]

| Depth (cm) | Field Size (cm×cm) | | | | | | | | | | | | | | | |
|---|---|---|---|---|---|---|---|---|---|---|---|---|---|---|---|---|
| | 4×4 | 5×5 | 6×6 | 8×8 | 10×10 | 12×12 | 15×15 | 18×18 | 20×20 | 22×22 | 25×25 | 30×30 | 32×32 | 35×35 | 38×38 | 40×40 |
| 1 | 1.156 | 1.140 | 1.129 | 1.116 | 1.107 | 1.100 | 1.092 | 1.087 | 1.083 | 1.081 | 1.077 | 1.071 | 1.070 | 1.067 | 1.064 | 1.062 |
| 2 | 1.136 | 1.123 | 1.115 | 1.102 | 1.094 | 1.088 | 1.082 | 1.077 | 1.073 | 1.071 | 1.067 | 1.062 | 1.060 | 1.058 | 1.056 | 1.054 |
| 3 | 1.095 | 1.086 | 1.080 | 1.073 | 1.069 | 1.066 | 1.061 | 1.057 | 1.055 | 1.052 | 1.050 | 1.048 | 1.046 | 1.044 | 1.042 | 1.041 |
| 4 | 1.049 | 1.046 | 1.043 | 1.039 | 1.037 | 1.034 | 1.032 | 1.029 | 1.028 | 1.027 | 1.026 | 1.023 | 1.022 | 1.022 | 1.021 | 1.021 |
| 5 | 1.000 | 1.000 | 1.000 | 1.000 | 1.000 | 1.000 | 1.000 | 1.000 | 1.000 | 1.000 | 1.000 | 1.000 | 1.000 | 1.000 | 1.000 | 1.000 |
| 6 | 0.959 | 0.961 | 0.963 | 0.966 | 0.968 | 0.970 | 0.972 | 0.973 | 0.974 | 0.975 | 0.976 | 0.977 | 0.978 | 0.978 | 0.979 | 0.979 |
| 8 | 0.864 | 0.870 | 0.877 | 0.887 | 0.896 | 0.903 | 0.912 | 0.918 | 0.920 | 0.923 | 0.927 | 0.930 | 0.932 | 0.933 | 0.934 | 0.936 |
| 10 | 0.784 | 0.794 | 0.802 | 0.817 | 0.829 | 0.838 | 0.850 | 0.859 | 0.863 | 0.868 | 0.872 | 0.879 | 0.881 | 0.884 | 0.888 | 0.888 |
| 12 | 0.703 | 0.716 | 0.727 | 0.744 | 0.760 | 0.772 | 0.787 | 0.798 | 0.804 | 0.810 | 0.817 | 0.826 | 0.829 | 0.832 | 0.836 | 0.837 |
| 14 | 0.631 | 0.645 | 0.657 | 0.678 | 0.694 | 0.709 | 0.725 | 0.738 | 0.745 | 0.752 | 0.760 | 0.771 | 0.774 | 0.780 | 0.784 | 0.787 |
| 17 | 0.534 | 0.549 | 0.561 | 0.582 | 0.601 | 0.618 | 0.648 | 0.654 | 0.663 | 0.671 | 0.681 | 0.694 | 0.699 | 0.705 | 0.710 | 0.712 |
| 20 | 0.456 | 0.469 | 0.481 | 0.501 | 0.518 | 0.533 | 0.555 | 0.572 | 0.583 | 0.592 | 0.603 | 0.619 | 0.625 | 0.631 | 0.638 | 0.642 |
| 24 | 0.370 | 0.378 | 0.388 | 0.408 | 0.424 | 0.440 | 0.460 | 0.479 | 0.489 | 0.499 | 0.512 | 0.529 | 0.536 | 0.543 | 0.551 | 0.555 |
| 28 | 0.299 | 0.307 | 0.314 | 0.332 | 0.348 | 0.362 | 0.380 | 0.397 | 0.406 | 0.414 | 0.427 | 0.444 | 0.450 | 0.460 | 0.468 | 0.473 |
| 32 | 0.245 | 0.249 | 0.256 | 0.270 | 0.282 | 0.296 | 0.312 | 0.327 | 0.337 | 0.345 | 0.356 | 0.372 | 0.379 | 0.388 | 0.395 | 0.400 |
| 36 | 0.200 | 0.203 | 0.208 | 0.220 | 0.231 | 0.242 | 0.258 | 0.272 | 0.280 | 0.288 | 0.298 | 0.313 | 0.319 | 0.327 | 0.332 | 0.339 |

From Biggs, P.J., Doppke, K.P., Leong, J.C., and Russell, M.D., Tissue phantom ratios for a Clinac 4/100, *Med. Phys.*, 9, 753, 1982. With permission.

## TABLE 1.C1
### Percent Depth Doses for 6-MV X Rays [Varian Clinac-6/100, 100 cm SSD]*

| Depth (cm) | Field Size (cm×cm) | | | | | | | | | | | | |
|---|---|---|---|---|---|---|---|---|---|---|---|---|
|  | 4×4 | 6×6 | 8×8 | 10×10 | 12×12 | 14×14 | 16×16 | 18×18 | 20×20 | 25×25 | 30×30 | 35×35 | 40×40 |
| 1.0 | 97.8 | 98.0 | 98.1 | 98.2 | 98.3 | 98.4 | 98.5 | 98.6 | 98.7 | 99.0 | 99.2 | 99.5 | 99.8 |
| $d_{max}$ | 100.0 | 100.0 | 100.0 | 100.0 | 100.0 | 100.0 | 100.0 | 100.0 | 100.0 | 100.0 | 100.0 | 100.0 | 100.0 |
| 2.0 | 98.1 | 98.5 | 98.5 | 98.5 | 98.5 | 98.5 | 98.5 | 98.5 | 98.5 | 98.5 | 98.5 | 98.5 | 98.5 |
| 3.0 | 93.5 | 94.0 | 94.4 | 94.7 | 94.8 | 94.9 | 94.9 | 94.9 | 95.0 | 95.0 | 95.0 | 95.1 | 95.2 |
| 4.0 | 88.5 | 89.4 | 90.0 | 90.4 | 90.8 | 91.0 | 91.3 | 91.4 | 91.5 | 91.6 | 91.6 | 91.7 | 91.8 |
| 5.0 | 83.5 | 85.0 | 86.1 | 86.8 | 87.2 | 87.4 | 87.5 | 87.6 | 87.7 | 87.9 | 88.2 | 88.4 | 88.7 |
| 6.0 | 79.3 | 80.8 | 81.9 | 82.6 | 83.2 | 83.5 | 83.7 | 83.8 | 84.0 | 84.2 | 84.6 | 84.8 | 85.2 |
| 7.0 | 74.4 | 76.4 | 77.8 | 78.6 | 79.2 | 79.6 | 79.8 | 80.1 | 80.2 | 80.6 | 81.1 | 81.5 | 82.0 |
| 8.0 | 70.1 | 72.0 | 73.6 | 74.6 | 75.4 | 75.8 | 76.2 | 76.5 | 76.8 | 77.3 | 77.7 | 78.0 | 78.5 |
| 9.0 | 65.3 | 67.6 | 69.4 | 70.6 | 71.4 | 72.0 | 72.6 | 73.0 | 73.3 | 73.8 | 74.3 | 74.7 | 75.2 |
| 10.0 | 61.9 | 64.1 | 65.6 | 66.8 | 67.8 | 68.5 | 69.0 | 69.6 | 70.0 | 70.6 | 71.0 | 71.6 | 72.0 |
| 11.0 | 57.6 | 60.2 | 62.0 | 63.4 | 64.4 | 65.2 | 65.8 | 66.2 | 66.7 | 67.2 | 67.7 | 68.2 | 68.8 |
| 12.0 | 53.9 | 56.2 | 58.2 | 59.6 | 60.6 | 61.4 | 62.0 | 62.6 | 63.0 | 64.0 | 64.5 | 65.1 | 65.6 |
| 13.0 | 50.6 | 52.8 | 55.0 | 56.6 | 57.8 | 58.6 | 59.2 | 59.7 | 60.1 | 60.9 | 61.6 | 62.3 | 63.0 |
| 14.0 | 47.5 | 49.8 | 51.8 | 53.4 | 54.6 | 55.5 | 56.2 | 56.8 | 57.2 | 58.0 | 58.8 | 59.4 | 60.1 |
| 15.0 | 44.7 | 46.7 | 48.6 | 50.2 | 51.6 | 52.6 | 53.4 | 54.0 | 54.4 | 55.4 | 56.1 | 56.7 | 57.3 |
| 16.0 | 41.7 | 43.7 | 45.6 | 47.4 | 48.8 | 49.8 | 50.6 | 51.4 | 51.9 | 53.1 | 53.8 | 54.2 | 54.5 |
| 17.0 | 39.2 | 41.2 | 43.0 | 44.5 | 45.8 | 47.0 | 47.8 | 48.6 | 49.2 | 50.4 | 51.2 | 51.8 | 52.2 |
| 18.0 | 36.8 | 38.6 | 40.4 | 41.9 | 43.2 | 44.4 | 45.3 | 46.0 | 46.8 | 48.0 | 49.0 | 49.5 | 50.0 |
| 19.0 | 34.6 | 36.4 | 38.4 | 39.9 | 41.2 | 42.2 | 43.1 | 43.8 | 44.4 | 45.6 | 46.5 | 47.1 | 47.6 |
| 20.0 | 32.6 | 34.5 | 36.2 | 37.7 | 38.9 | 39.8 | 40.7 | 41.4 | 42.0 | 43.3 | 44.1 | 44.8 | 45.4 |
| 22.0 | 28.8 | 30.6 | 32.2 | 33.7 | 35.0 | 36.0 | 36.8 | 37.6 | 38.1 | 39.2 | 40.1 | 40.8 | 41.6 |
| 24.0 | 25.3 | 27.0 | 28.6 | 30.0 | 31.2 | 32.2 | 33.0 | 33.6 | 34.2 | 35.5 | 36.4 | 37.0 | 37.7 |
| 26.0 | 22.4 | 23.9 | 25.2 | 26.4 | 27.6 | 28.6 | 29.4 | 30.2 | 30.8 | 32.1 | 33.0 | 33.5 | 34.1 |
| 28.0 | 19.9 | 21.3 | 22.6 | 23.7 | 24.8 | 25.6 | 26.5 | 27.2 | 27.9 | 29.1 | 29.8 | 30.4 | 31.0 |
| 30.0 | 17.5 | 18.8 | 20.0 | 21.1 | 22.2 | 23.0 | 23.8 | 24.4 | 25.0 | 26.2 | 27.0 | 27.6 | 28.2 |

* CL-600C units exhibit similar depth dose values. See Ref. 13.

From Coffey, II, C.W., Beach, J.L., Thompson, D.J., and Mendiondo, M., X-ray beam characteristics of the Varian Clinac 6-100 linear accelerator, *Med. Phys.*, 7, 716, 1980. With permission.

## TABLE 1.C2
### Tissue Phantom Ratios for 6-MV X Rays [Varian Clinac-6/100, 100 cm SAD]*

| Depth (cm) | Field Size (cm×cm) | | | | | | | | | |
|---|---|---|---|---|---|---|---|---|---|---|
| | 0×0 | 4×4 | 6×6 | 8×8 | 10×10 | 12×12 | 16×16 | 20×20 | 30×30 | 40×40 |
| 1.0 | 0.950 | 0.966 | 0.968 | 0.969 | 0.970 | 0.972 | 0.974 | 0.977 | 0.983 | 0.990 |
| 1.5 | 0.926 | 1.000 | 1.000 | 1.000 | 1.000 | 1.000 | 1.000 | 1.000 | 1.000 | 1.000 |
| 2.0 | 0.903 | 0.991 | 0.992 | 0.992 | 0.992 | 0.993 | 0.994 | 0.995 | 0.996 | 0.998 |
| 3.0 | 0.858 | 0.957 | 0.964 | 0.969 | 0.972 | 0.974 | 0.976 | 0.977 | 0.981 | 0.984 |
| 4.0 | 0.815 | 0.924 | 0.934 | 0.944 | 0.949 | 0.953 | 0.958 | 0.959 | 0.961 | 0.965 |
| 5.0 | 0.775 | 0.891 | 0.904 | 0.914 | 0.921 | 0.926 | 0.933 | 0.936 | 0.942 | 0.949 |
| 6.0 | 0.736 | 0.858 | 0.874 | 0.886 | 0.894 | 0.901 | 0.910 | 0.915 | 0.924 | 0.930 |
| 7.0 | 0.699 | 0.822 | 0.841 | 0.856 | 0.866 | 0.874 | 0.885 | 0.892 | 0.901 | 0.909 |
| 8.0 | 0.665 | 0.787 | 0.806 | 0.822 | 0.834 | 0.842 | 0.858 | 0.868 | 0.880 | 0.887 |
| 9.0 | 0.632 | 0.749 | 0.769 | 0.786 | 0.802 | 0.814 | 0.832 | 0.842 | 0.856 | 0.865 |
| 10.0 | 0.600 | 0.720 | 0.740 | 0.758 | 0.774 | 0.788 | 0.808 | 0.821 | 0.834 | 0.846 |
| 12.0 | 0.542 | 0.649 | 0.672 | 0.694 | 0.712 | 0.728 | 0.748 | 0.762 | 0.781 | 0.797 |
| 14.0 | 0.489 | 0.590 | 0.614 | 0.636 | 0.656 | 0.672 | 0.698 | 0.714 | 0.736 | 0.754 |
| 15.0 | 0.465 | 0.560 | 0.584 | 0.608 | 0.628 | 0.646 | 0.675 | 0.692 | 0.714 | 0.731 |
| 16.0 | 0.442 | 0.535 | 0.559 | 0.581 | 0.601 | 0.620 | 0.648 | 0.667 | 0.693 | 0.712 |
| 18.0 | 0.399 | 0.479 | 0.504 | 0.526 | 0.547 | 0.566 | 0.596 | 0.617 | 0.647 | 0.668 |
| 20.0 | 0.360 | 0.439 | 0.464 | 0.486 | 0.507 | 0.525 | 0.554 | 0.574 | 0.603 | 0.625 |
| 22.0 | 0.325 | 0.399 | 0.422 | 0.444 | 0.464 | 0.482 | 0.511 | 0.531 | 0.564 | 0.586 |
| 24.0 | 0.294 | 0.359 | 0.384 | 0.406 | 0.426 | 0.444 | 0.472 | 0.492 | 0.530 | 0.551 |
| 26.0 | 0.265 | 0.328 | 0.350 | 0.371 | 0.390 | 0.407 | 0.434 | 0.454 | 0.488 | 0.514 |
| 28.0 | 0.240 | 0.297 | 0.319 | 0.338 | 0.356 | 0.373 | 0.398 | 0.420 | 0.454 | 0.480 |
| 30.0 | 0.216 | 0.270 | 0.290 | 0.308 | 0.326 | 0.342 | 0.366 | 0.386 | 0.423 | 0.446 |

* CL-600C units exhibit similar depth dose values. See Ref. 13.

From Coffey, II, C.W., Beach, J.L., Thompson, D.J., and Mendiondo, M., X-ray beam characteristics of the Varian Clinac 6-100 linear accelerator, *Med. Phys.*, 7, 716, 1980. With permission.

## TABLE 1.C3
### Percent Depth Doses for 6-MV X Rays [Varian Clinac-2500, 100 cm SSD]*

| Depth (cm) | Field Size (cm×cm) | | | | | | | | | | | |
|---|---|---|---|---|---|---|---|---|---|---|---|---|
| | 4×4 | 5×5 | 6×6 | 8×8 | 10×10 | 12×12 | 15×15 | 18×18 | 20×20 | 25×25 | 30×30 | 35×35 |
| 1.5 | 100.0 | 100.0 | 100.0 | 100.0 | 100.0 | 100.0 | 100.0 | 100.0 | 100.0 | 100.0 | 100.0 | 100.0 |
| 2.0 | 98.1 | 98.2 | 98.3 | 98.5 | 98.6 | 98.6 | 98.7 | 98.8 | 98.9 | 99.0 | 99.1 | 99.4 |
| 3.0 | 93.6 | 93.9 | 94.1 | 94.5 | 94.8 | 95.0 | 95.3 | 95.6 | 95.7 | 96.0 | 96.3 | 96.5 |
| 4.0 | 88.8 | 89.3 | 89.7 | 90.3 | 90.7 | 91.0 | 91.6 | 92.0 | 92.2 | 92.7 | 93.1 | 93.5 |
| 5.0 | 84.8 | 85.4 | 85.9 | 86.7 | 87.3 | 87.8 | 88.4 | 88.9 | 89.2 | 89.9 | 90.4 | 90.8 |
| 6.0 | 79.6 | 80.3 | 81.0 | 81.9 | 82.7 | 83.4 | 84.1 | 84.8 | 85.2 | 86.0 | 86.6 | 87.2 |
| 7.0 | 74.8 | 75.6 | 76.3 | 77.5 | 78.4 | 79.1 | 80.1 | 80.8 | 81.3 | 82.2 | 83.0 | 83.6 |
| 8.0 | 70.2 | 71.2 | 72.0 | 73.3 | 74.3 | 75.1 | 76.2 | 77.0 | 77.5 | 78.6 | 79.5 | 80.2 |
| 9.0 | 66.0 | 67.1 | 67.9 | 69.3 | 70.4 | 71.3 | 72.4 | 73.4 | 73.9 | 75.1 | 76.1 | 76.9 |
| 10.0 | 62.1 | 63.2 | 64.1 | 65.6 | 66.7 | 67.7 | 68.9 | 69.9 | 70.5 | 71.7 | 72.8 | 73.7 |
| 11.0 | 58.4 | 59.5 | 60.5 | 62.0 | 63.2 | 64.2 | 65.5 | 66.6 | 67.2 | 68.5 | 69.6 | 70.6 |
| 12.0 | 54.9 | 56.1 | 57.1 | 58.7 | 59.9 | 61.0 | 62.3 | 63.4 | 64.0 | 65.4 | 66.6 | 67.6 |
| 13.0 | 51.6 | 52.8 | 53.9 | 55.5 | 56.8 | 57.9 | 59.2 | 60.4 | 61.0 | 62.4 | 63.6 | 64.7 |
| 14.0 | 48.6 | 49.8 | 50.8 | 52.5 | 53.8 | 54.9 | 56.3 | 57.4 | 58.1 | 59.6 | 60.8 | 61.8 |
| 15.0 | 45.7 | 47.0 | 48.0 | 49.7 | 51.0 | 52.1 | 53.5 | 54.7 | 55.4 | 56.8 | 58.1 | 59.2 |
| 16.0 | 43.1 | 44.3 | 45.3 | 47.0 | 48.3 | 49.4 | 50.8 | 52.0 | 52.7 | 54.2 | 55.5 | 56.6 |
| 17.0 | 40.6 | 41.8 | 42.8 | 44.5 | 45.8 | 46.9 | 48.3 | 49.5 | 50.2 | 51.7 | 53.0 | 54.0 |
| 18.0 | 38.2 | 39.4 | 40.4 | 42.1 | 43.4 | 44.5 | 45.9 | 47.1 | 47.8 | 49.3 | 50.5 | 51.6 |
| 19.0 | 36.0 | 37.2 | 38.2 | 39.8 | 41.1 | 42.2 | 43.6 | 44.8 | 45.5 | 46.9 | 49.2 | 49.3 |
| 20.0 | 33.9 | 35.1 | 36.1 | 37.7 | 39.0 | 40.0 | 41.4 | 42.6 | 43.2 | 44.7 | 46.0 | 47.0 |
| 21.0 | 32.0 | 33.1 | 34.1 | 35.6 | 36.9 | 38.0 | 39.3 | 40.5 | 41.1 | 42.6 | 43.8 | 44.9 |
| 22.0 | 30.2 | 31.3 | 32.2 | 33.7 | 35.0 | 36.0 | 37.3 | 38.5 | 39.1 | 40.5 | 41.8 | 42.8 |
| 23.0 | 28.4 | 29.5 | 30.4 | 31.9 | 33.1 | 34.2 | 35.4 | 36.5 | 37.2 | 38.6 | 39.8 | 40.8 |
| 24.0 | 26.8 | 27.9 | 28.8 | 30.2 | 31.4 | 32.4 | 33.7 | 34.7 | 35.3 | 36.7 | 37.9 | 38.9 |
| 25.0 | 25.3 | 26.3 | 27.2 | 28.6 | 29.7 | 30.7 | 31.9 | 33.0 | 33.6 | 34.9 | 36.1 | 37.1 |
| 26.0 | 23.9 | 24.9 | 25.7 | 27.1 | 28.2 | 29.1 | 30.3 | 31.3 | 31.9 | 33.2 | 34.3 | 35.3 |

*Clinac-1800 units exhibit similar depth dose values. See Ref. 18.

Adapted from The University of Iowa Hospitals and Clinics, Iowa City, IA, 1992, personal communication.

**TABLE 1.C4**

**Tissue Phantom Ratios for 6-MV X Rays [Varian Clinac-2500, 100 cm SAD]\***

| Depth (cm) | Field Size (cm×cm) | | | | | | | | | | | | |
|---|---|---|---|---|---|---|---|---|---|---|---|---|---|
| | 0×0 | 4×4 | 5×5 | 6×6 | 8×8 | 10×10 | 12×12 | 15×15 | 18×18 | 20×20 | 25×25 | 30×30 | 35×35 |
| 1.5 | 1.000 | 1.000 | 1.000 | 1.000 | 1.000 | 1.000 | 1.000 | 1.000 | 1.000 | 1.000 | 1.000 | 1.000 | 1.000 |
| 2.0 | 0.977 | 0.983 | 0.984 | 0.985 | 0.987 | 0.988 | 0.990 | 0.992 | 0.994 | 0.994 | 0.994 | 0.995 | 0.996 |
| 3.0 | 0.932 | 0.956 | 0.960 | 0.963 | 0.967 | 0.970 | 0.971 | 0.972 | 0.976 | 0.976 | 0.976 | 0.977 | 0.978 |
| 4.0 | 0.889 | 0.923 | 0.929 | 0.934 | 0.940 | 0.945 | 0.948 | 0.951 | 0.955 | 0.955 | 0.956 | 0.957 | 0.960 |
| 5.0 | 0.848 | 0.889 | 0.895 | 0.902 | 0.910 | 0.917 | 0.922 | 0.926 | 0.930 | 0.931 | 0.934 | 0.936 | 0.941 |
| 6.0 | 0.809 | 0.856 | 0.864 | 0.870 | 0.882 | 0.891 | 0.897 | 0.903 | 0.910 | 0.910 | 0.915 | 0.918 | 0.922 |
| 7.0 | 0.772 | 0.823 | 0.832 | 0.847 | 0.853 | 0.862 | 0.870 | 0.877 | 0.885 | 0.885 | 0.893 | 0.900 | 0.904 |
| 8.0 | 0.737 | 0.787 | 0.796 | 0.805 | 0.820 | 0.832 | 0.842 | 0.852 | 0.864 | 0.864 | 0.782 | 0.877 | 0.880 |
| 9.0 | 0.703 | 0.756 | 0.765 | 0.774 | 0.788 | 0.800 | 0.811 | 0.823 | 0.839 | 0.839 | 0.849 | 0.854 | 0.858 |
| 10.0 | 0.671 | 0.721 | 0.731 | 0.740 | 0.756 | 0.770 | 0.782 | 0.796 | 0.814 | 0.814 | 0.827 | 0.832 | 0.836 |
| 11.0 | 0.640 | 0.691 | 0.701 | 0.712 | 0.724 | 0.743 | 0.756 | 0.775 | 0.791 | 0.791 | 0.803 | 0.809 | 0.815 |
| 12.0 | 0.610 | 0.666 | 0.676 | 0.686 | 0.704 | 0.719 | 0.732 | 0.747 | 0.767 | 0.767 | 0.778 | 0.788 | 0.794 |
| 13.0 | 0.582 | 0.637 | 0.648 | 0.658 | 0.675 | 0.690 | 0.702 | 0.718 | 0.741 | 0.741 | 0.757 | 0.765 | 0.773 |
| 14.0 | 0.556 | 0.607 | 0.614 | 0.628 | 0.647 | 0.663 | 0.676 | 0.694 | 0.716 | 0.716 | 0.730 | 0.743 | 0.751 |
| 15.0 | 0.530 | 0.582 | 0.593 | 0.604 | 0.623 | 0.639 | 0.653 | 0.670 | 0.691 | 0.691 | 0.706 | 0.719 | 0.729 |
| 16.0 | 0.506 | 0.556 | 0.566 | 0.576 | 0.594 | 0.612 | 0.626 | 0.642 | 0.669 | 0.669 | 0.686 | 0.699 | 0.708 |
| 17.0 | 0.483 | 0.531 | 0.542 | 0.552 | 0.571 | 0.587 | 0.603 | 0.622 | 0.647 | 0.647 | 0.666 | 0.680 | 0.688 |
| 18.0 | 0.460 | 0.508 | 0.519 | 0.529 | 0.549 | 0.564 | 0.589 | 0.602 | 0.626 | 0.626 | 0.644 | 0.657 | 0.666 |
| 19.0 | 0.439 | 0.484 | 0.499 | 0.509 | 0.527 | 0.544 | 0.560 | 0.580 | 0.602 | 0.602 | 0.621 | 0.635 | 0.648 |
| 20.0 | 0.419 | 0.466 | 0.476 | 0.487 | 0.506 | 0.522 | 0.537 | 0.557 | 0.582 | 0.582 | 0.600 | 0.615 | 0.628 |
| 21.0 | 0.400 | 0.447 | 0.458 | 0.467 | 0.486 | 0.502 | 0.518 | 0.537 | 0.561 | 0.561 | 0.580 | 0.596 | 0.607 |
| 22.0 | 0.381 | 0.426 | 0.436 | 0.446 | 0.464 | 0.480 | 0.494 | 0.514 | 0.542 | 0.542 | 0.560 | 0.575 | 0.587 |
| 23.0 | 0.364 | 0.408 | 0.418 | 0.428 | 0.446 | 0.463 | 0.477 | 0.497 | 0.522 | 0.522 | 0.542 | 0.557 | 0.568 |
| 24.0 | 0.347 | 0.391 | 0.400 | 0.410 | 0.426 | 0.444 | 0.458 | 0.479 | 0.504 | 0.504 | 0.524 | 0.538 | 0.549 |
| 25.0 | 0.331 | 0.373 | 0.382 | 0.391 | 0.409 | 0.425 | 0.440 | 0.459 | 0.485 | 0.485 | 0.505 | 0.517 | 0.530 |

\*Clinac-1800 units exhibit similar depth dose values. See Ref. 18.
Adapted from The University of Iowa Hospitals and Clinics, Iowa City, IA, 1992, personal communication.

## TABLE 1.C5
### Percent Depth Doses for 6-MV X Rays [Siemens Mevatron-67, 100 cm SSD]

| Depth (cm) | \multicolumn{15}{c}{Field Size (cm×cm)} |
|---|

| Depth (cm) | 4×4 | 5×5 | 6×6 | 8×8 | 10×10 | 12×12 | 14×14 | 15×15 | 16×16 | 18×18 | 20×20 | 25×25 | 30×30 | 35×35 | 40×40 |
|---|---|---|---|---|---|---|---|---|---|---|---|---|---|---|---|
| 1.5 | 100.0 | 100.0 | 100.0 | 100.0 | 100.0 | 100.0 | 100.0 | 100.0 | 100.0 | 100.0 | 100.0 | 100.0 | 100.0 | 100.0 | 100.0 |
| 2.0 | 98.2 | 98.3 | 98.3 | 98.5 | 98.6 | 98.7 | 98.7 | 98.8 | 98.8 | 98.8 | 98.8 | 98.8 | 98.9 | 98.9 | 98.9 |
| 3.0 | 93.9 | 94.1 | 94.3 | 94.6 | 94.9 | 95.2 | 95.4 | 95.5 | 95.5 | 95.6 | 95.7 | 95.8 | 95.8 | 95.9 | 96.0 |
| 4.0 | 89.4 | 89.7 | 90.0 | 90.5 | 91.0 | 91.5 | 91.8 | 91.9 | 92.0 | 92.1 | 92.2 | 92.4 | 92.5 | 92.7 | 92.9 |
| 5.0 | 85.0 | 85.3 | 85.6 | 86.4 | 87.1 | 87.6 | 88.0 | 88.2 | 88.3 | 88.5 | 88.7 | 88.9 | 89.1 | 89.4 | 89.7 |
| 6.0 | 80.6 | 81.0 | 81.4 | 82.3 | 83.1 | 83.8 | 84.3 | 84.5 | 84.7 | 85.0 | 85.1 | 85.4 | 85.7 | 86.0 | 86.4 |
| 8.0 | 72.3 | 72.8 | 73.3 | 74.4 | 75.5 | 76.4 | 77.1 | 77.3 | 77.6 | 77.9 | 78.2 | 78.6 | 79.0 | 79.5 | 80.0 |
| 10.0 | 64.7 | 65.2 | 65.8 | 67.1 | 68.3 | 69.4 | 70.2 | 70.5 | 70.8 | 71.2 | 71.5 | 72.1 | 72.6 | 73.2 | 73.9 |
| 12.0 | 57.7 | 58.3 | 59.0 | 60.4 | 61.7 | 62.9 | 63.8 | 64.1 | 64.5 | 64.9 | 65.3 | 65.9 | 66.6 | 67.3 | 68.1 |
| 14.0 | 51.4 | 52.1 | 52.7 | 54.2 | 55.6 | 56.9 | 57.9 | 58.2 | 58.6 | 59.1 | 59.5 | 60.2 | 60.9 | 61.7 | 62.7 |
| 16.0 | 45.8 | 46.4 | 47.1 | 48.6 | 50.1 | 51.4 | 52.4 | 52.8 | 53.2 | 53.7 | 54.1 | 54.9 | 55.7 | 56.6 | 57.6 |
| 18.0 | 40.7 | 41.4 | 42.1 | 43.6 | 45.1 | 46.4 | 47.4 | 47.8 | 48.2 | 48.8 | 49.2 | 50.0 | 50.8 | 51.8 | 52.9 |
| 20.0 | 36.2 | 36.8 | 37.5 | 39.0 | 40.5 | 41.8 | 42.8 | 43.3 | 43.6 | 44.2 | 44.7 | 45.5 | 46.4 | 47.4 | 48.5 |
| 22.0 | 32.2 | 32.8 | 33.4 | 34.9 | 36.4 | 37.6 | 38.7 | 39.1 | 39.5 | 40.1 | 40.5 | 41.4 | 42.3 | 43.3 | 44.4 |
| 24.0 | 28.6 | 29.1 | 29.8 | 31.2 | 32.6 | 33.9 | 34.9 | 35.3 | 35.7 | 36.3 | 36.8 | 37.6 | 38.5 | 39.5 | 40.7 |
| 26.0 | 25.4 | 25.9 | 26.5 | 27.9 | 29.3 | 30.5 | 31.5 | 31.9 | 32.3 | 32.9 | 33.3 | 34.2 | 35.1 | 36.1 | 37.3 |
| 28.0 | 22.5 | 23.0 | 23.6 | 24.9 | 26.2 | 27.4 | 28.4 | 28.8 | 29.2 | 29.7 | 30.2 | 31.0 | 31.9 | 32.9 | 34.1 |
| 30.0 | 20.0 | 20.5 | 21.0 | 22.3 | 23.5 | 24.7 | 25.6 | 26.0 | 26.3 | 26.9 | 27.3 | 28.2 | 29.0 | 30.0 | 31.2 |
| 32.0 | 17.7 | 18.2 | 18.7 | 19.9 | 21.1 | 22.2 | 23.1 | 23.4 | 23.8 | 24.3 | 24.7 | 25.6 | 26.4 | 27.4 | 28.6 |
| 34.0 | 15.7 | 16.1 | 16.6 | 17.7 | 18.9 | 19.9 | 20.8 | 21.1 | 21.5 | 22.0 | 22.4 | 23.2 | 24.0 | 25.0 | 26.1 |

From Horton, J.L., Dosimetry of the Siemens mevatron 67 linear accelerator, *Int. J. Radiat. Oncol. Biol. Phys.*, 9, 1217, 1983. With permission.

## TABLE 1.C6
### Tissue Phantom Ratios for 6-MV X Rays [Siemens Mevatron-67, 100 cm SAD]

| Depth (cm) | Field Size (cm×cm) | | | | | | | |
|---|---|---|---|---|---|---|---|---|
| | 5×5 | 10×10 | 15×15 | 20×20 | 25×25 | 30×30 | 35×35 | 40×40 |
| 0.37 | 0.668 | 0.695 | 0.721 | 0.747 | 0.768 | 0.795 | 0.810 | 0.821 |
| 0.75 | 0.851 | 0.870 | 0.885 | 0.901 | 0.915 | 0.929 | 0.938 | 0.943 |
| 1.12 | 0.962 | 0.967 | 0.969 | 0.979 | 0.983 | 0.985 | 0.990 | 0.995 |
| 1.50 | 1.000 | 1.000 | 1.000 | 1.000 | 1.000 | 1.000 | 1.000 | 1.000 |
| 2.00 | 1.011 | 1.009 | 1.009 | 1.009 | 1.009 | 1.008 | 1.007 | 1.007 |
| 3.00 | 0.983 | 0.987 | 0.989 | 0.990 | 0.993 | 0.994 | 0.993 | 0.992 |
| 4.00 | 0.949 | 0.963 | 0.968 | 0.970 | 0.976 | 0.978 | 0.977 | 0.976 |
| 5.00 | 0.915 | 0.935 | 0.944 | 0.949 | 0.958 | 0.964 | 0.967 | 0.967 |
| 6.00 | 0.884 | 0.907 | 0.918 | 0.930 | 0.941 | 0.946 | 0.952 | 0.952 |
| 8.00 | 0.826 | 0.857 | 0.877 | 0.890 | 0.898 | 0.907 | 0.910 | 0.910 |
| 10.00 | 0.757 | 0.801 | 0.822 | 0.836 | 0.843 | 0.853 | 0.860 | 0.862 |
| 12.00 | 0.696 | 0.744 | 0.771 | 0.789 | 0.802 | 0.815 | 0.822 | 0.825 |
| 14.00 | 0.641 | 0.691 | 0.723 | 0.744 | 0.761 | 0.775 | 0.783 | 0.788 |
| 16.00 | 0.595 | 0.645 | 0.681 | 0.704 | 0.722 | 0.736 | 0.744 | 0.744 |
| 18.00 | 0.546 | 0.595 | 0.633 | 0.658 | 0.677 | 0.692 | 0.702 | 0.706 |
| 20.00 | 0.504 | 0.552 | 0.587 | 0.614 | 0.636 | 0.653 | 0.666 | 0.671 |
| 22.00 | 0.465 | 0.512 | 0.549 | 0.576 | 0.597 | 0.616 | 0.629 | 0.635 |
| 24.00 | 0.428 | 0.474 | 0.509 | 0.538 | 0.560 | 0.579 | 0.592 | 0.598 |
| 26.00 | 0.392 | 0.437 | 0.472 | 0.501 | 0.523 | 0.543 | 0.558 | 0.565 |
| 28.00 | 0.363 | 0.405 | 0.439 | 0.466 | 0.489 | 0.510 | 0.524 | 0.529 |
| 30.00 | 0.334 | 0.374 | 0.407 | 0.433 | 0.458 | 0.477 | 0.491 | 0.498 |

From Horton, J.L., Dosimetry of the Siemens mevatron 67 linear accelerator, *Int. J. Radiat. Oncol. Biol. Phys.*, 9, 1217, 1983. With permission.

## TABLE 1.C7
## Percent Depth Doses for 6-MV X Rays [Siemens Mevatron KD, 100 cm SSD]

| Depth (cm) | Field Size (cm×cm) | | | | | | | | | | | | |
|---|---|---|---|---|---|---|---|---|---|---|---|---|---|
| | 0×0 | 2×2 | 3×3 | 4×4 | 5×5 | 8×8 | 10×10 | 12×12 | 15×15 | 20×20 | 25×25 | 30×30 | 35×35 |
| 0.0 | 13.1 | 15.6 | 16.8 | 18.0 | 19.2 | 22.4 | 24.4 | 25.7 | 27.4 | 30.3 | 32.0 | 33.3 | 34.6 |
| 1.0 | 96.6 | 96.3 | 96.2 | 96.0 | 95.9 | 96.1 | 96.4 | 96.7 | 97.1 | 97.9 | 98.5 | 99.0 | 99.6 |
| 2.0 | 100.0 | 100.0 | 100.0 | 100.0 | 100.0 | 100.0 | 100.0 | 100.0 | 100.0 | 100.0 | 100.0 | 100.0 | 100.0 |
| 3.0 | 94.5 | 94.8 | 94.8 | 94.9 | 95.2 | 95.8 | 96.0 | 96.1 | 96.2 | 96.3 | 96.0 | 95.5 | 94.9 |
| 4.0 | 86.9 | 88.2 | 89.0 | 89.7 | 90.2 | 91.4 | 91.8 | 92.1 | 92.3 | 92.6 | 92.8 | 92.8 | 92.7 |
| 5.0 | 81.2 | 83.1 | 84.4 | 85.3 | 86.0 | 87.4 | 88.0 | 88.3 | 88.7 | 89.1 | 89.5 | 89.7 | 89.8 |
| 6.0 | 76.4 | 78.3 | 79.5 | 80.4 | 81.2 | 82.9 | 83.7 | 84.2 | 84.7 | 85.3 | 85.9 | 86.2 | 86.4 |
| 7.0 | 71.8 | 73.7 | 74.8 | 75.7 | 76.5 | 78.6 | 79.6 | 80.2 | 80.9 | 81.7 | 82.4 | 82.8 | 83.2 |
| 8.0 | 66.1 | 68.6 | 70.1 | 71.3 | 72.4 | 74.6 | 75.7 | 76.4 | 77.2 | 78.1 | 78.9 | 79.4 | 79.8 |
| 10.0 | 55.2 | 59.0 | 61.3 | 63.1 | 64.5 | 67.0 | 68.2 | 69.1 | 70.2 | 71.4 | 72.3 | 72.9 | 73.4 |
| 15.0 | 39.7 | 43.2 | 45.1 | 46.7 | 48.0 | 50.7 | 52.1 | 53.3 | 54.6 | 56.1 | 57.0 | 57.3 | 57.6 |
| 20.0 | 29.3 | 32.1 | 33.7 | 34.8 | 35.4 | 38.4 | 40.0 | 41.1 | 42.5 | 44.3 | 45.6 | 46.6 | 47.4 |
| 25.0 | 21.6 | 24.0 | 25.3 | 26.4 | 27.3 | 29.4 | 30.6 | 31.6 | 32.9 | 34.8 | 36.2 | 37.5 | 38.6 |
| 30.0 | 14.9 | 17.0 | 18.0 | 19.2 | 20.3 | 21.5 | 22.4 | 23.5 | 24.7 | 26.6 | 28.2 | 29.7 | 31.0 |

From Al-Ghazi, M.S.A.L., Arjune, B., Fiedler, J.A., and Sharma, P.D., Dosimetric aspects of the therapeutic photon beams from a dual-energy linear accelerator, *Med. Phys.*, 15, 250, 1988. With permission.

## TABLE 1.C8
### Tissue Phantom Ratios for 6-MV X Rays [Siemens Mevatron KD, 100 cm SAD]

| Depth (cm) | Field Size (cm×cm) | | | | | | | | | | | | |
|---|---|---|---|---|---|---|---|---|---|---|---|---|
| | 0×0 | 2×2 | 3×3 | 4×4 | 5×5 | 8×6 | 10×10 | 12×12 | 15×15 | 20×20 | 25×25 | 30×30 | 35×35 |
| 0.0 | 0.126 | 0.150 | 0.162 | 0.173 | 0.185 | 0.215 | 0.235 | 0.247 | 0.263 | 0.291 | 0.308 | 0.321 | 0.333 |
| 1.0 | 0.947 | 0.944 | 0.943 | 0.942 | 0.941 | 0.942 | 0.945 | 0.948 | 0.952 | 0.960 | 0.966 | 0.971 | 0.977 |
| 2.0 | 1.000 | 1.000 | 1.000 | 1.000 | 1.000 | 1.000 | 1.000 | 1.000 | 1.000 | 1.000 | 1.000 | 1.000 | 1.000 |
| 3.0 | 0.964 | 0.967 | 0.967 | 0.968 | 0.970 | 0.976 | 0.978 | 0.979 | 0.981 | 0.981 | 0.979 | 0.974 | 0.968 |
| 4.0 | 0.903 | 0.916 | 0.924 | 0.931 | 0.937 | 0.949 | 0.954 | 0.956 | 0.959 | 0.962 | 0.963 | 0.963 | 0.962 |
| 5.0 | 0.860 | 0.879 | 0.892 | 0.902 | 0.910 | 0.924 | 0.930 | 0.934 | 0.938 | 0.943 | 0.946 | 0.947 | 0.948 |
| 6.0 | 0.825 | 0.844 | 0.856 | 0.866 | 0.874 | 0.893 | 0.901 | 0.907 | 0.913 | 0.919 | 0.924 | 0.927 | 0.929 |
| 7.0 | 0.791 | 0.810 | 0.821 | 0.830 | 0.839 | 0.862 | 0.872 | 0.879 | 0.887 | 0.896 | 0.902 | 0.906 | 0.910 |
| 8.0 | 0.741 | 0.767 | 0.782 | 0.796 | 0.807 | 0.832 | 0.844 | 0.852 | 0.861 | 0.872 | 0.879 | 0.884 | 0.888 |
| 10.0 | 0.641 | 0.681 | 0.706 | 0.727 | 0.743 | 0.773 | 0.787 | 0.797 | 0.810 | 0.824 | 0.833 | 0.839 | 0.844 |
| 15.0 | 0.505 | 0.543 | 0.564 | 0.584 | 0.600 | 0.634 | 0.651 | 0.665 | 0.682 | 0.702 | 0.714 | 0.719 | 0.721 |
| 20.0 | 0.405 | 0.437 | 0.455 | 0.472 | 0.484 | 0.512 | 0.535 | 0.552 | 0.571 | 0.545 | 0.613 | 0.624 | 0.633 |
| 25.0 | 0.324 | 0.353 | 0.368 | 0.383 | 0.396 | 0.425 | 0.441 | 0.454 | 0.473 | 0.499 | 0.518 | 0.533 | 0.546 |
| 30.0 | 0.243 | 0.268 | 0.281 | 0.294 | 0.308 | 0.338 | 0.346 | 0.357 | 0.376 | 0.402 | 0.424 | 0.442 | 0.458 |

From Al-Ghazi, M.S.A.L., Arjune, B., Fiedler, J.A., and Sharma, P.D., Dosimetric aspects of the therapeutic photon beams from a dual-energy linear accelerator, *Med. Phys.*, 15, 250, 1988. With permission.

**TABLE 1.C9**

**Percent Depth Doses for 6-MV X Rays [Philips SL-25, 100 cm SSD]**

| Depth (cm) | Field Size (cm×cm) | | | | | | | | | | |
|---|---|---|---|---|---|---|---|---|---|---|---|
| | 4×4 | 6×6 | 8×8 | 10×10 | 12×12 | 15×15 | 20×20 | 25×25 | 30×30 | 35×35 | 40×40 |
| 1.0 | 94.1 | 94.6 | 95.0 | 95.6 | 96.1 | 96.6 | 97.4 | 97.8 | 98.1 | 98.3 | 98.2 |
| 1.5 | 100.0 | 100.0 | 100.0 | 100.0 | 100.0 | 100.0 | 100.0 | 100.0 | 100.0 | 100.0 | 100.0 |
| 2.0 | 99.2 | 99.3 | 99.4 | 99.4 | 99.3 | 99.0 | 98.8 | 98.8 | 98.9 | 98.9 | 99.0 |
| 3.0 | 94.9 | 95.3 | 95.4 | 95.5 | 95.6 | 95.5 | 95.4 | 95.4 | 95.5 | 95.6 | 95.8 |
| 4.0 | 89.8 | 90.7 | 91.0 | 91.3 | 91.4 | 91.6 | 91.7 | 91.9 | 92.0 | 92.1 | 92.3 |
| 5.0 | 84.8 | 86.1 | 86.6 | 87.0 | 87.2 | 87.5 | 87.9 | 88.3 | 88.5 | 88.6 | 88.7 |
| 6.0 | 80.3 | 81.6 | 82.5 | 83.0 | 83.4 | 83.8 | 84.2 | 84.7 | 84.9 | 85.2 | 85.3 |
| 7.0 | 76.0 | 77.5 | 78.5 | 79.2 | 79.7 | 80.2 | 80.8 | 81.4 | 81.7 | 82.0 | 82.2 |
| 8.0 | 71.8 | 73.5 | 74.6 | 75.5 | 76.0 | 76.7 | 77.5 | 78.2 | 78.6 | 79.0 | 79.2 |
| 9.0 | 67.7 | 69.5 | 70.7 | 71.7 | 72.4 | 73.1 | 74.0 | 74.8 | 75.3 | 75.7 | 75.9 |
| 10.0 | 63.8 | 65.7 | 67.0 | 68.1 | 68.9 | 69.7 | 70.7 | 71.5 | 72.1 | 72.5 | 72.8 |
| 11.0 | 60.1 | 62.0 | 64.4 | 64.6 | 65.5 | 66.4 | 67.5 | 68.4 | 69.1 | 69.5 | 69.7 |
| 12.0 | 56.6 | 58.5 | 60.0 | 61.2 | 62.0 | 63.2 | 64.5 | 65.4 | 66.1 | 66.5 | 66.8 |
| 13.0 | 53.5 | 55.4 | 56.8 | 58.0 | 59.0 | 60.2 | 61.5 | 62.5 | 63.2 | 63.6 | 63.8 |
| 14.0 | 50.5 | 52.3 | 53.8 | 55.0 | 56.0 | 57.3 | 58.7 | 59.7 | 60.4 | 60.9 | 61.1 |
| 15.0 | 47.6 | 49.3 | 50.8 | 52.2 | 53.3 | 54.5 | 56.0 | 57.0 | 57.7 | 58.2 | 58.4 |
| 16.0 | 44.7 | 46.4 | 48.0 | 49.3 | 50.5 | 51.7 | 53.2 | 54.3 | 55.1 | 55.5 | 55.8 |
| 17.0 | 41.8 | 43.7 | 45.3 | 46.7 | 47.8 | 49.0 | 50.5 | 51.7 | 52.5 | 53.0 | 53.2 |
| 18.0 | 39.6 | 41.3 | 42.9 | 44.2 | 45.4 | 46.5 | 48.2 | 49.3 | 50.2 | 50.7 | 50.7 |
| 19.0 | 37.5 | 39.0 | 40.6 | 41.9 | 43.0 | 44.3 | 45.9 | 47.0 | 47.9 | 48.3 | 48.4 |
| 20.0 | 35.4 | 36.9 | 38.3 | 39.7 | 40.7 | 42.0 | 43.5 | 44.8 | 45.7 | 46.1 | 46.2 |
| 21.0 | 33.5 | 34.9 | 36.3 | 37.5 | 38.6 | 39.9 | 41.5 | 42.8 | 43.5 | 44.0 | 44.0 |
| 22.0 | 31.6 | 33.0 | 34.3 | 35.7 | 36.6 | 37.9 | 39.5 | 40.8 | 41.5 | 41.9 | 42.0 |
| 23.0 | 30.0 | 31.2 | 32.5 | 33.8 | 34.8 | 36.0 | 37.7 | 38.9 | 39.7 | 40.0 | 40.1 |
| 24.0 | 28.3 | 29.5 | 30.8 | 32.0 | 33.0 | 34.3 | 35.9 | 37.1 | 37.9 | 38.2 | 38.2 |
| 25.0 | 26.7 | 27.9 | 29.1 | 30.3 | 31.3 | 32.5 | 34.1 | 35.4 | 36.1 | 36.4 | 36.4 |
| 26.0 | 25.2 | 26.4 | 27.6 | 28.8 | 29.7 | 30.9 | 32.5 | 33.7 | 34.4 | 34.7 | 34.7 |
| 28.0 | 22.4 | 23.7 | 24.8 | 25.8 | 26.8 | 27.9 | 29.4 | 30.6 | 31.2 | 31.5 | 31.4 |
| 30.0 | 19.9 | 21.1 | 22.1 | 23.1 | 24.0 | 25.1 | 26.5 | 27.6 | 28.3 | 28.5 | 28.5 |
| 32.0 | 17.9 | 19.0 | 20.0 | 20.9 | 21.7 | 22.8 | 24.1 | 25.1 | 25.8 | 25.9 | 26.0 |
| 34.0 | 16.0 | 17.0 | 17.9 | 18.8 | 19.6 | 20.5 | 21.8 | 22.8 | 23.4 | 23.5 | 23.5 |

From Palta, J.R., Ayyangar, K.M., and Suntharalingam, N., Dosimetric characteristics of a 6 MV photon beam from a linear accelerator with asymmetric collimator jaws, *Int. J. Radiat. Oncol. Biol. Phys.*, 14, 383, 1988. With permission.

## TABLE 1.C10
## Tissue Phantom Ratios for 6-MV X Rays [Philips SL-25, 100 cm SAD]

| Depth (cm) | Field Size (cm×cm) | | | | | | | | | | |
|---|---|---|---|---|---|---|---|---|---|---|---|
| | 4×4 | 6×6 | 8×8 | 10×10 | 12×12 | 15×15 | 20×20 | 25×25 | 30×30 | 35×35 | 40×40 |
| 1.0 | 0.932 | 0.936 | 0.941 | 0.947 | 0.951 | 0.956 | 0.964 | 0.969 | 0.972 | 0.973 | 0.973 |
| 1.5 | 1.000 | 1.000 | 1.000 | 1.000 | 1.000 | 1.000 | 1.000 | 1.000 | 1.000 | 1.000 | 1.000 |
| 2.0 | 1.001 | 1.003 | 1.003 | 1.004 | 1.003 | 1.000 | 0.998 | 0.998 | 0.998 | 0.999 | 1.000 |
| 3.0 | 0.976 | 0.980 | 0.982 | 0.983 | 0.984 | 0.983 | 0.982 | 0.982 | 0.983 | 0.984 | 0.986 |
| 4.0 | 0.942 | 0.951 | 0.954 | 0.957 | 0.959 | 0.960 | 0.961 | 0.963 | 0.965 | 0.966 | 0.968 |
| 5.0 | 0.905 | 0.919 | 0.925 | 0.929 | 0.932 | 0.935 | 0.939 | 0.942 | 0.945 | 0.947 | 0.949 |
| 6.0 | 0.872 | 0.887 | 0.897 | 0.903 | 0.907 | 0.912 | 0.916 | 0.921 | 0.924 | 0.928 | 0.930 |
| 7.0 | 0.840 | 0.858 | 0.868 | 0.877 | 0.882 | 0.888 | 0.895 | 0.900 | 0.905 | 0.909 | 0.912 |
| 8.0 | 0.809 | 0.827 | 0.840 | 0.849 | 0.857 | 0.863 | 0.872 | 0.880 | 0.886 | 0.891 | 0.895 |
| 9.0 | 0.776 | 0.796 | 0.810 | 0.821 | 0.830 | 0.838 | 0.848 | 0.857 | 0.865 | 0.870 | 0.874 |
| 10.0 | 0.743 | 0.764 | 0.780 | 0.792 | 0.802 | 0.812 | 0.824 | 0.833 | 0.842 | 0.848 | 0.853 |
| 11.0 | 0.713 | 0.734 | 0.751 | 0.764 | 0.775 | 0.786 | 0.800 | 0.811 | 0.820 | 0.826 | 0.831 |
| 12.0 | 0.682 | 0.704 | 0.722 | 0.736 | 0.747 | 0.761 | 0.776 | 0.788 | 0.798 | 0.805 | 0.810 |
| 13.0 | 0.655 | 0.677 | 0.694 | 0.710 | 0.721 | 0.736 | 0.752 | 0.766 | 0.775 | 0.783 | 0.789 |
| 14.0 | 0.629 | 0.650 | 0.667 | 0.683 | 0.696 | 0.711 | 0.729 | 0.743 | 0.754 | 0.762 | 0.767 |
| 15.0 | 0.603 | 0.623 | 0.640 | 0.656 | 0.670 | 0.687 | 0.706 | 0.720 | 0.732 | 0.740 | 0.746 |
| 16.0 | 0.575 | 0.595 | 0.613 | 0.631 | 0.645 | 0.662 | 0.682 | 0.697 | 0.709 | 0.718 | 0.725 |
| 17.0 | 0.546 | 0.568 | 0.587 | 0.604 | 0.619 | 0.637 | 0.658 | 0.674 | 0.687 | 0.697 | 0.703 |
| 18.0 | 0.526 | 0.547 | 0.565 | 0.582 | 0.597 | 0.614 | 0.636 | 0.652 | 0.666 | 0.676 | 0.683 |
| 19.0 | 0.506 | 0.525 | 0.543 | 0.559 | 0.574 | 0.592 | 0.613 | 0.631 | 0.645 | 0.655 | 0.663 |
| 20.0 | 0.486 | 0.503 | 0.520 | 0.537 | 0.551 | 0.569 | 0.591 | 0.609 | 0.624 | 0.635 | 0.642 |
| 21.0 | 0.467 | 0.484 | 0.500 | 0.516 | 0.531 | 0.549 | 0.571 | 0.589 | 0.604 | 0.615 | 0.622 |
| 22.0 | 0.448 | 0.464 | 0.479 | 0.495 | 0.511 | 0.528 | 0.551 | 0.570 | 0.585 | 0.596 | 0.603 |
| 23.0 | 0.431 | 0.446 | 0.460 | 0.477 | 0.492 | 0.509 | 0.533 | 0.552 | 0.567 | 0.577 | 0.585 |
| 24.0 | 0.414 | 0.428 | 0.442 | 0.458 | 0.472 | 0.490 | 0.513 | 0.533 | 0.548 | 0.559 | 0.567 |
| 25.0 | 0.397 | 0.410 | 0.423 | 0.439 | 0.453 | 0.470 | 0.495 | 0.515 | 0.530 | 0.541 | 0.549 |
| 26.0 | 0.379 | 0.394 | 0.407 | 0.423 | 0.436 | 0.454 | 0.477 | 0.497 | 0.512 | 0.524 | 0.531 |
| 27.0 | 0.361 | 0.377 | 0.391 | 0.406 | 0.420 | 0.437 | 0.460 | 0.479 | 0.495 | 0.506 | 0.513 |
| 28.0 | 0.345 | 0.361 | 0.375 | 0.390 | 0.404 | 0.420 | 0.443 | 0.462 | 0.477 | 0.489 | 0.497 |
| 29.0 | 0.330 | 0.346 | 0.360 | 0.375 | 0.387 | 0.404 | 0.427 | 0.446 | 0.461 | 0.472 | 0.480 |
| 30.0 | 0.314 | 0.330 | 0.344 | 0.359 | 0.371 | 0.387 | 0.410 | 0.429 | 0.444 | 0.455 | 0.463 |
| 32.0 | 0.291 | 0.305 | 0.319 | 0.332 | 0.344 | 0.360 | 0.382 | 0.400 | 0.415 | 0.426 | 0.435 |
| 34.0 | 0.267 | 0.281 | 0.293 | 0.306 | 0.317 | 0.333 | 0.354 | 0.371 | 0.387 | 0.398 | 0.405 |

From Palta, J.R., Ayyangar, K.M., and Suntharalingam, N., Dosimetric characteristics of a 6 MV photon beam from a linear accelerator with asymmetric collimator jaws, *Int. J. Radiat. Oncol. Biol. Phys.*, 14, 383, 1988. With permission.

**TABLE 1.C11**

**Percent Depth Doses for 6-MV X Rays [Scanditronix MM-22, 100 cm SSD]**

| Depth (cm) | Field Size (cm×cm) | | | | | | | | | | | |
|---|---|---|---|---|---|---|---|---|---|---|---|---|
| | 4×4 | 6×6 | 8×8 | 10×10 | 12×12 | 14×14 | 16×16 | 18×18 | 20×20 | 25×25 | 30×30 | 35×35 |
| 1.6 | 100.0 | 100.0 | 100.0 | 100.0 | 100.0 | 100.0 | 100.0 | 100.0 | 100.0 | 100.0 | 100.0 | 100.0 |
| 2.0 | 98.9 | 99.8 | 100.0 | 100.0 | 100.0 | 100.0 | 100.0 | 100.0 | 100.0 | 100.0 | 99.7 | 99.2 |
| 3.0 | 94.3 | 96.2 | 96.5 | 96.5 | 96.8 | 97.0 | 96.6 | 96.6 | 96.6 | 96.6 | 96.4 | 96.4 |
| 4.0 | 89.5 | 91.6 | 92.3 | 92.4 | 92.5 | 92.5 | 92.5 | 92.5 | 92.5 | 92.8 | 93.0 | 93.4 |
| 5.0 | 84.2 | 86.6 | 87.6 | 87.9 | 88.3 | 89.0 | 88.6 | 89.0 | 89.1 | 89.2 | 89.2 | 89.5 |
| 6.0 | 79.2 | 81.9 | 83.4 | 83.9 | 83.9 | 84.9 | 85.3 | 85.4 | 85.7 | 85.7 | 85.7 | 86.0 |
| 7.0 | 74.5 | 76.8 | 79.2 | 79.5 | 80.0 | 81.2 | 81.3 | 81.6 | 81.9 | 82.2 | 82.3 | 82.5 |
| 8.0 | 70.2 | 73.2 | 75.0 | 75.2 | 76.4 | 77.4 | 77.4 | 77.7 | 78.5 | 78.7 | 79.1 | 79.7 |
| 9.0 | 66.0 | 69.5 | 70.9 | 71.8 | 72.7 | 74.0 | 74.0 | 74.6 | 74.8 | 75.4 | 75.8 | 76.1 |
| 10.0 | 61.9 | 65.0 | 67.4 | 68.4 | 69.1 | 70.3 | 70.3 | 70.7 | 71.9 | 72.2 | 72.4 | 73.3 |
| 12.0 | 54.7 | 58.1 | 60.4 | 61.4 | 62.3 | 63.8 | 64.1 | 65.0 | 65.1 | 66.0 | 66.4 | 66.9 |
| 14.0 | 48.5 | 51.8 | 53.8 | 54.9 | 56.1 | 57.3 | 57.9 | 58.4 | 59.4 | 60.4 | 60.8 | 61.6 |
| 16.0 | 43.0 | 46.1 | 47.8 | 48.9 | 50.3 | 51.8 | 52.2 | 53.1 | 54.0 | 55.2 | 55.6 | 56.8 |
| 18.0 | 38.1 | 40.8 | 43.0 | 44.1 | 45.3 | 46.8 | 47.4 | 48.2 | 49.3 | 50.2 | 50.8 | 51.6 |
| 20.0 | 33.5 | 36.5 | 38.5 | 39.7 | 40.5 | 41.9 | 42.6 | 43.4 | 44.5 | 45.9 | 46.4 | 47.1 |
| 22.0 | 30.1 | 32.7 | 34.5 | 35.2 | 36.5 | 37.8 | 38.6 | 39.0 | 40.5 | 41.7 | 42.5 | 43.0 |
| 24.0 | 26.9 | 28.9 | 30.7 | 31.6 | 32.6 | 34.0 | 34.7 | 35.3 | 36.9 | 38.1 | 38.9 | 39.6 |
| 26.0 | 23.8 | 25.7 | 27.3 | 28.3 | 29.5 | 30.5 | 31.5 | 32.2 | 33.3 | 34.7 | 35.3 | 36.0 |
| 28.0 | 21.2 | 23.0 | 24.6 | 25.9 | 26.6 | 27.6 | 28.4 | 28.8 | 30.0 | 31.5 | 32.1 | 33.0 |
| 30.0 | 19.0 | 20.7 | 22.1 | 23.0 | 23.8 | 25.1 | 25.6 | 26.2 | 27.6 | 28.8 | 29.5 | 30.4 |

From George, R.E., Characteristics of an MM 22 medical microtron 6-MV photon beam, *Med. Phys.*, 11, 862, 1984. With permission.

## TABLE 1.C12
### Tissue Phantom Ratios for 6-MV X Rays [Scanditronix MM-22, 100 cm SAD]

| Depth (cm) | Field Size (cm×cm) | | | | | | | | | | | | |
|---|---|---|---|---|---|---|---|---|---|---|---|---|---|
| | 0×0 | 4×4 | 6×6 | 8×8 | 10×10 | 12×12 | 14×14 | 16×16 | 18×18 | 20×20 | 25×25 | 30×30 | 35×35 |
| 1.6 | 1.000 | 1.000 | 1.000 | 1.000 | 1.000 | 1.000 | 1.000 | 1.000 | 1.000 | 1.000 | 1.000 | 1.000 | 1.000 |
| 2.0 | 0.982 | 0.996 | 1.000 | 1.000 | 1.000 | 1.000 | 1.000 | 1.000 | 1.000 | 1.000 | 1.000 | 1.000 | 1.000 |
| 3.0 | 0.938 | 0.969 | 0.987 | 0.994 | 0.994 | 0.997 | 0.997 | 0.997 | 0.997 | 0.997 | 0.995 | 0.994 | 0.994 |
| 4.0 | 0.895 | 0.938 | 0.961 | 0.968 | 0.970 | 0.971 | 0.974 | 0.972 | 0.971 | 0.973 | 0.974 | 0.977 | 0.979 |
| 5.0 | 0.855 | 0.899 | 0.924 | 0.935 | 0.941 | 0.944 | 0.950 | 0.952 | 0.952 | 0.952 | 0.954 | 0.956 | 0.957 |
| 6.0 | 0.817 | 0.862 | 0.890 | 0.905 | 0.914 | 0.915 | 0.925 | 0.927 | 0.931 | 0.930 | 0.934 | 0.936 | 0.938 |
| 7.0 | 0.780 | 0.826 | 0.855 | 0.874 | 0.882 | 0.887 | 0.898 | 0.902 | 0.905 | 0.908 | 0.912 | 0.915 | 0.917 |
| 8.0 | 0.745 | 0.793 | 0.823 | 0.842 | 0.852 | 0.859 | 0.872 | 0.876 | 0.881 | 0.885 | 0.891 | 0.895 | 0.898 |
| 9.0 | 0.711 | 0.760 | 0.793 | 0.812 | 0.824 | 0.834 | 0.847 | 0.851 | 0.857 | 0.862 | 0.870 | 0.875 | 0.879 |
| 10.0 | 0.679 | 0.726 | 0.762 | 0.783 | 0.798 | 0.807 | 0.821 | 0.826 | 0.832 | 0.840 | 0.848 | 0.854 | 0.857 |
| 12.0 | 0.620 | 0.665 | 0.702 | 0.722 | 0.742 | 0.752 | 0.766 | 0.777 | 0.782 | 0.790 | 0.804 | 0.811 | 0.814 |
| 14.0 | 0.565 | 0.611 | 0.647 | 0.667 | 0.683 | 0.698 | 0.714 | 0.723 | 0.731 | 0.740 | 0.757 | 0.763 | 0.772 |
| 16.0 | 0.516 | 0.561 | 0.592 | 0.611 | 0.629 | 0.644 | 0.663 | 0.674 | 0.684 | 0.692 | 0.712 | 0.724 | 0.732 |
| 18.0 | 0.470 | 0.514 | 0.542 | 0.562 | 0.585 | 0.598 | 0.616 | 0.629 | 0.639 | 0.649 | 0.671 | 0.680 | 0.691 |
| 20.0 | 0.429 | 0.467 | 0.498 | 0.520 | 0.541 | 0.555 | 0.568 | 0.582 | 0.592 | 0.602 | 0.626 | 0.640 | 0.649 |
| 22.0 | 0.391 | 0.434 | 0.461 | 0.480 | 0.500 | 0.506 | 0.527 | 0.540 | 0.549 | 0.558 | 0.585 | 0.601 | 0.612 |
| 24.0 | 0.357 | 0.401 | 0.419 | 0.439 | 0.460 | 0.468 | 0.488 | 0.500 | 0.510 | 0.521 | 0.552 | 0.565 | 0.576 |
| 26.0 | 0.325 | 0.366 | 0.380 | 0.401 | 0.424 | 0.433 | 0.452 | 0.463 | 0.475 | 0.487 | 0.513 | 0.530 | 0.542 |
| 28.0 | 0.297 | 0.336 | 0.351 | 0.369 | 0.390 | 0.402 | 0.419 | 0.430 | 0.442 | 0.452 | 0.477 | 0.496 | 0.509 |
| 30.0 | 0.271 | 0.311 | 0.325 | 0.341 | 0.359 | 0.372 | 0.385 | 0.399 | 0.410 | 0.419 | 0.448 | 0.466 | 0.478 |

From George, R.E., Characteristics of an MM 22 medical microtron 6-MV photon beam, *Med. Phys.*, 11, 862, 1984. With permission.

## TABLE 1.C13
## Percent Depth Doses for 6-MV X Rays [Mitsubishi EXL-8, 100 cm SSD]

| Depth (cm) | Field Size (cm×cm) | | | | | | | |
|---|---|---|---|---|---|---|---|---|
| | 3×3 | 5×5 | 10×10 | 15×15 | 20×20 | 25×25 | 30×30 | 35×35 |
| 0.0 | 11.4 | 13.4 | 18.2 | 22.5 | 28.1 | 30.4 | 33.4 | 36.5 |
| 0.5 | 79.4 | 80.2 | 79.4 | 82.1 | 83.8 | 84.8 | 84.7 | 85.0 |
| 1.0 | 96.4 | 96.3 | 96.3 | 97.6 | 98.1 | 98.4 | 98.1 | 98.4 |
| 1.5 | 100.0 | 100.0 | 100.0 | 100.0 | 100.0 | 100.0 | 100.0 | 100.0 |
| 2.0 | 98.5 | 98.5 | 98.9 | 98.3 | 98.4 | 98.4 | 98.1 | 98.2 |
| 3.0 | 93.9 | 94.4 | 95.1 | 94.8 | 94.9 | 95.1 | 95.1 | 95.1 |
| 4.0 | 88.4 | 89.7 | 90.8 | 91.1 | 91.2 | 91.4 | 91.4 | 91.7 |
| 5.0 | 83.2 | 85.0 | 86.8 | 87.2 | 87.4 | 87.8 | 87.9 | 88.0 |
| 6.0 | 78.4 | 80.3 | 82.7 | 83.3 | 83.7 | 84.2 | 84.3 | 84.6 |
| 7.0 | 73.5 | 76.0 | 78.7 | 79.7 | 80.1 | 80.7 | 80.9 | 81.1 |
| 8.0 | 69.0 | 71.8 | 74.9 | 76.0 | 76.7 | 77.3 | 77.6 | 77.9 |
| 9.0 | 64.8 | 67.8 | 71.2 | 72.4 | 73.3 | 73.9 | 74.2 | 74.7 |
| 10.0 | 60.8 | 63.7 | 67.5 | 69.0 | 70.0 | 70.7 | 71.1 | 71.6 |
| 11.0 | 57.0 | 60.1 | 64.1 | 65.6 | 66.8 | 67.6 | 67.9 | 68.4 |
| 12.0 | 53.5 | 56.6 | 60.6 | 62.5 | 63.7 | 64.6 | 65.0 | 65.6 |
| 13.0 | 50.3 | 53.5 | 57.6 | 59.4 | 60.7 | 61.6 | 62.1 | 62.7 |
| 14.0 | 47.0 | 50.1 | 54.5 | 56.1 | 57.7 | 58.7 | 59.4 | 59.9 |
| 15.0 | 44.0 | 47.3 | 51.7 | 53.3 | 55.0 | 56.1 | 56.3 | 56.7 |
| 16.0 | 41.1 | 44.2 | 48.8 | 50.4 | 52.1 | 53.0 | 53.7 | 54.4 |
| 18.0 | 36.5 | 39.3 | 43.9 | 45.7 | 47.5 | 48.5 | 49.3 | 49.7 |
| 20.0 | 32.3 | 35.0 | 39.4 | 41.3 | 42.9 | 44.1 | 44.7 | 45.3 |
| 22.0 | 28.8 | 31.4 | 35.3 | 37.4 | 39.1 | 40.2 | 41.0 | 41.6 |
| 24.0 | 25.6 | 28.0 | 31.7 | 33.8 | 35.5 | 36.6 | 37.4 | 37.9 |
| 26.0 | 22.8 | 24.9 | 28.5 | 30.6 | 32.1 | 33.2 | 33.9 | 34.6 |
| 28.0 | 20.3 | 22.2 | 25.6 | 27.6 | 29.1 | 30.2 | 30.9 | 31.4 |
| 30.0 | 18.5 | 20.0 | 23.0 | 25.1 | 26.8 | 27.8 | 28.2 | 29.1 |

From Sharma, S.C., Modur, P., and Basavatia, R., Evaluation of a photon and an electron beam of a 6-MV linear accelerator, *Med. Phys.*, 15, 525, 1988. With permission.

**TABLE 1.C14**

**Tissue Phantom Ratios for 6-MV X Rays [Mitsubishi EXL-8 , 100 cm SAD]**

| Depth (cm) | Field Size (cm×cm) | | | | | | | |
|---|---|---|---|---|---|---|---|---|
| | 3×3 | 5×5 | 10×10 | 15×15 | 20×20 | 25×25 | 30×30 | 35×35 |
| 0.0 | 0.111 | 0.130 | 0.177 | 0.218 | 0.273 | 0.295 | 0.324 | 0.355 |
| 0.5 | 0.778 | 0.786 | 0.779 | 0.804 | 0.822 | 0.831 | 0.831 | 0.834 |
| 1.0 | 0.954 | 0.953 | 0.954 | 0.966 | 0.971 | 0.974 | 0.972 | 0.974 |
| 1.5 | 1.000 | 1.000 | 1.000 | 1.000 | 1.000 | 1.000 | 1.000 | 1.000 |
| 2.0 | 0.995 | 0.995 | 0.998 | 0.993 | 0.994 | 0.994 | 0.991 | 0.992 |
| 3.0 | 0.967 | 0.972 | 0.979 | 0.977 | 0.978 | 0.979 | 0.978 | 0.979 |
| 4.0 | 0.928 | 0.941 | 0.952 | 0.956 | 0.957 | 0.959 | 0.960 | 0.962 |
| 5.0 | 0.890 | 0.908 | 0.927 | 0.933 | 0.935 | 0.939 | 0.940 | 0.941 |
| 6.0 | 0.854 | 0.873 | 0.899 | 0.908 | 0.912 | 0.917 | 0.919 | 0.922 |
| 7.0 | 0.815 | 0.841 | 0.871 | 0.883 | 0.889 | 0.895 | 0.898 | 0.900 |
| 8.0 | 0.779 | 0.808 | 0.843 | 0.858 | 0.866 | 0.873 | 0.877 | 0.880 |
| 9.0 | 0.744 | 0.776 | 0.814 | 0.832 | 0.842 | 0.849 | 0.854 | 0.858 |
| 10.0 | 0.710 | 0.742 | 0.785 | 0.806 | 0.817 | 0.826 | 0.832 | 0.837 |
| 11.0 | 0.678 | 0.712 | 0.757 | 0.779 | 0.793 | 0.803 | 0.810 | 0.814 |
| 12.0 | 0.648 | 0.681 | 0.728 | 0.753 | 0.769 | 0.781 | 0.789 | 0.793 |
| 13.0 | 0.618 | 0.654 | 0.702 | 0.728 | 0.745 | 0.757 | 0.766 | 0.771 |
| 14.0 | 0.588 | 0.623 | 0.674 | 0.700 | 0.718 | 0.732 | 0.743 | 0.751 |
| 15.0 | 0.559 | 0.596 | 0.649 | 0.676 | 0.695 | 0.711 | 0.721 | 0.723 |
| 16.0 | 0.531 | 0.566 | 0.621 | 0.650 | 0.668 | 0.684 | 0.694 | 0.702 |
| 18.0 | 0.488 | 0.520 | 0.574 | 0.606 | 0.627 | 0.645 | 0.657 | 0.665 |
| 20.0 | 0.445 | 0.477 | 0.530 | 0.564 | 0.585 | 0.603 | 0.616 | 0.623 |
| 22.0 | 0.409 | 0.440 | 0.490 | 0.524 | 0.547 | 0.566 | 0.578 | 0.589 |
| 24.0 | 0.375 | 0.404 | 0.452 | 0.486 | 0.510 | 0.530 | 0.543 | 0.553 |
| 26.0 | 0.344 | 0.370 | 0.416 | 0.451 | 0.475 | 0.494 | 0.508 | 0.518 |
| 28.0 | 0.316 | 0.339 | 0.383 | 0.418 | 0.442 | 0.461 | 0.475 | 0.485 |
| 30.0 | 0.298 | 0.317 | 0.355 | 0.388 | 0.414 | 0.436 | 0.449 | 0.458 |

From Sharma, S.C., Modur, P., and Basavatia, R., Evaluation of a photon and an electron beam of a 6-MV linear accelerator, *Med. Phys.*, 15, 525, 1988. With permission.

## TABLE 1.C15
## Percent Depth Doses for 6-MV X Rays [AECL Therac-6, 100 cm SSD]

| Depth (cm) | Field Size (cm×cm) | | | | | | | | | |
|---|---|---|---|---|---|---|---|---|---|---|
| | 0×0 | 4×4 | 6×6 | 8×8 | 10×10 | 15×15 | 20×20 | 30×30 | 40×40 |
| 1.5 | 100.0 | 100.0 | 100.0 | 100.0 | 100.0 | 100.0 | 100.0 | 100.0 | 100.0 |
| 2.0 | 97.5 | 97.9 | 98.2 | 98.4 | 98.5 | 98.6 | 98.6 | 98.8 | 99.0 |
| 3.0 | 91.5 | 93.2 | 93.9 | 94.4 | 94.6 | 94.9 | 95.0 | 95.5 | 96.0 |
| 4.0 | 85.9 | 88.4 | 89.4 | 90.1 | 90.5 | 91.0 | 91.4 | 91.6 | 92.5 |
| 5.0 | 80.7 | 83.5 | 85.0 | 85.9 | 86.4 | 87.1 | 87.5 | 88.0 | 88.6 |
| 6.0 | 75.7 | 79.0 | 80.5 | 81.7 | 82.4 | 83.2 | 83.6 | 84.4 | 85.0 |
| 7.0 | 71.1 | 74.5 | 76.4 | 77.5 | 78.4 | 79.4 | 79.9 | 80.7 | 81.4 |
| 8.0 | 66.7 | 70.4 | 72.2 | 73.5 | 74.5 | 75.6 | 76.4 | 77.4 | 78.2 |
| 9.0 | 62.6 | 66.4 | 68.4 | 69.7 | 70.7 | 72.0 | 72.7 | 73.9 | 75.0 |
| 10.0 | 58.7 | 62.5 | 64.5 | 66.0 | 67.1 | 68.6 | 69.4 | 70.5 | 72.2 |
| 12.0 | 51.6 | 55.5 | 57.6 | 59.3 | 60.4 | 62.0 | 62.9 | 64.3 | 66.0 |
| 14.0 | 45.3 | 49.2 | 51.4 | 53.1 | 54.3 | 56.1 | 57.0 | 58.9 | 60.6 |
| 16.0 | 39.8 | 43.5 | 45.7 | 47.5 | 48.7 | 50.6 | 51.6 | 53.6 | 55.4 |
| 18.0 | 35.0 | 38.5 | 40.6 | 42.4 | 43.7 | 45.6 | 46.6 | 48.7 | 50.6 |
| 20.0 | 30.8 | 34.1 | 36.0 | 37.6 | 39.0 | 41.0 | 42.1 | 44.2 | 46.2 |
| 22.0 | 27.0 | 30.2 | 32.0 | 33.7 | 35.0 | 37.0 | 38.0 | 40.1 | 42.0 |
| 24.0 | 23.7 | 26.6 | 28.5 | 30.1 | 31.4 | 33.3 | 34.3 | 36.4 | 38.4 |
| 26.0 | 20.9 | 23.6 | 25.3 | 26.8 | 28.0 | 30.0 | 31.0 | 33.0 | 34.9 |
| 28.0 | 18.3 | 20.9 | 22.5 | 23.9 | 25.0 | 26.9 | 28.0 | 30.0 | 31.9 |
| 30.0 | 16.1 | 18.5 | 19.9 | 21.1 | 22.2 | 24.2 | 25.0 | 27.8 | 29.1 |

From Grant III, W., Ames, J., and Almond, P.R., Evaluation of the Therac 6 linear accelerator for radiation therapy. *Med. Phys.*, 5, 448, 1978. With permission.

## TABLE 1.D1
### Tissue Phantom Ratios for 8-MV X Rays [ARCO Mevatron VIII, 100 cm SAD]

| Depth (cm) | Field Size (cm×cm) | | | | | | | | | | |
|---|---|---|---|---|---|---|---|---|---|---|---|
| | 4×4 | 5×5 | 6×6 | 7×7 | 8×8 | 10×10 | 12×12 | 15×15 | 17×17 | 20×20 | 25×25 |
| 1.0 | 0.907 | 0.911 | 0.913 | 0.918 | 0.920 | 0.926 | 0.930 | 0.931 | 0.935 | 0.940 | 0.951 |
| 2.0 | 1.000 | 1.000 | 1.000 | 1.000 | 1.000 | 1.000 | 1.000 | 1.000 | 1.000 | 1.000 | 1.000 |
| 3.0. | 0.983 | 0.986 | 0.989 | 0.991 | 0.994 | 0.995 | 0.995 | 0.995 | 0.995 | 0.995 | 0.995 |
| 4.0 | 0.952 | 0.960 | 0.963 | 0.968 | 0.971 | 0.974 | 0.976 | 0.977 | 0.977 | 0.978 | 0.979 |
| 5.0 | 0.920 | 0.926 | 0.928 | 0.933 | 0.938 | 0.944 | 0.950 | 0.951 | 0.952 | 0.953 | 0.955 |
| 6.0 | 0.881 | 0.887 | 0.895 | 0.901 | 0.907 | 0.914 | 0.923 | 0.930 | 0.933 | 0.935 | 0.939 |
| 7.0 | 0.849 | 0.855 | 0.862 | 0.870 | 0.873 | 0.881 | 0.890 | 0.900 | 0.903 | 0.910 | 0.917 |
| 8.0 | 0.810 | 0.820 | 0.823 | 0.836 | 0.846 | 0.860 | 0.871 | 0.876 | 0.879 | 0.886 | 0.890 |
| 10.0 | 0.754 | 0.761 | 0.768 | 0.775 | 0.782 | 0.799 | 0.810 | 0.823 | 0.830 | 0.837 | 0.847 |
| 12.0 | 0.700 | 0.708 | 0.714 | 0.726 | 0.737 | 0.752 | 0.766 | 0.779 | 0.785 | 0.791 | 0.806 |
| 14.0 | 0.650 | 0.657 | 0.663 | 0.670 | 0.682 | 0.700 | 0.716 | 0.731 | 0.738 | 0.748 | 0.768 |
| 16.0 | 0.598 | 0.605 | 0.612 | 0.620 | 0.630 | 0.650 | 0.667 | 0.682 | 0.691 | 0.703 | 0.720 |
| 18.0 | 0.549 | 0.555 | 0.563 | 0.572 | 0.581 | 0.600 | 0.619 | 0.639 | 0.651 | 0.665 | 0.682 |
| 20.0 | 0.514 | 0.521 | 0.527 | 0.535 | 0.543 | 0.562 | 0.578 | 0.598 | 0.606 | 0.619 | 0.636 |

From Agarwal, S.K., Scheele. R.V., and Wakley, J., Tissue maximum-dose ratio (TMR) for 8 MV x rays, *Am. J. Roentgenol. Rad. Ther. Nucl. Med.*, 112, 797, 1971. With permission.

## TABLE 1.D2
### Percent Depth Doses for 9-MV X Rays [CGR Neptune-10, 100 cm SSD]

| Depth (cm) | Field Size (cm×cm) | | | | | | | |
|---|---|---|---|---|---|---|---|---|
| | 0×0 | 1×1 | 2×2 | 4×4 | 5×5 | 10×10 | 20×20 | 30×30 |
| 0.5 | (73.0)* | (73.5) | (74.5) | 76.0 | 76.5 | 79.0 | 85.0 | 89.0 |
| 1.0 | (90.0) | (91.0) | (92.0) | 92.0 | 92.5 | 93.5 | 96.0 | 98.0 |
| 1.3 | 100.0 | 99.5 | 99.5 | 97.0 | 97.7 | 98.0 | 98.5 | 99.5 |
| 1.5 | 98.8 | 99.9 | 99.7 | 99.0 | 99.0 | 99.0 | 99.0 | 99.9 |
| 1.6 | 98.3 | 100.0 | 99.9 | 99.5 | 99.5 | 99.5 | 100.0 | 100.0 |
| 1.8 | 97.1 | 99.3 | 100.0 | 99.7 | 99.8 | 99.5 | 100.0 | 99.5 |
| 1.9 | 96.6 | 98.8 | 99.9 | 100.0 | 99.9 | 100.0 | 99.9 | 99.5 |
| 2.0 | 96.0 | 98.0 | 98.9 | 100.0 | 100.0 | 99.9 | 99.5 | 99.2 |
| 5.0 | 80.7 | 84.0 | 85.0 | 88.0 | 89.1 | 89.5 | 90.5 | 90.0 |
| 10.0 | 60.9 | 62.0 | 64.1 | 67.5 | 68.5 | 71.0 | 73.0 | 73.5 |
| 15.0 | 46.1 | 46.0 | 48.6 | 51.0 | 52.5 | 55.5 | 59.0 | 59.5 |
| 20.0 | 35.0 | 35.0 | 36.8 | 38.5 | 40.0 | 43.0 | 47.0 | 48.1 |
| 25.0 | 26.7 | 26.7 | 27.5 | 29.5 | 30.7 | 34.0 | 37.5 | 38.0 |
| 30.0 | 20.4 | 20.4 | 21.5 | 22.5 | 24.0 | 27.0 | 29.5 | 30.6 |

*Parentheses indicate data obtained by extrapolation.

From Arcovito, G., Piermattei, A., D'Abramo, G., and Bassi, F.A., Dose measurements and calculations of small radiation fields for 9-MV x rays, *Med. Phys.*, 12, 779, 1985. With permission.

**TABLE 1.D3**

**Percent Depth Doses for 10-MV X Rays [Varian Clinac-18, 100 cm SSD]**

| Depth (cm) | Field Size (cm×cm) | | | | | | | | | | |
|---|---|---|---|---|---|---|---|---|---|---|---|
| | 4×4 | 5×5 | 6×6 | 8×8 | 10×10 | 12×12 | 15×15 | 20×20 | 25×25 | 30×30 | 35×35 |
| 1.0 | 89.4 | 90.1 | 90.6 | 91.3 | 92.3 | 93.0 | 94.0 | 96.1 | 97.4 | 98.0 | 98.3 |
| 2.0 | 100.1 | 100.1 | 100.1 | 100.1 | 100.5 | 100.7 | 100.8 | 101.0 | 101.0 | 100.9 | 100.8 |
| 2.5 | 100.0 | 100.0 | 100.0 | 100.0 | 100.0 | 100.0 | 100.0 | 100.0 | 100.0 | 100.0 | 100.0 |
| 3.0 | 98.3 | 98.3 | 98.3 | 98.4 | 98.4 | 98.3 | 98.1 | 98.1 | 97.9 | 98.1 | 98.3 |
| 4.0 | 93.9 | 94.3 | 94.4 | 94.7 | 95.1 | 95.2 | 95.1 | 94.9 | 95.0 | 95.1 | 95.3 |
| 5.0 | 90.9 | 90.4 | 90.7 | 91.1 | 91.4 | 91.6 | 91.8 | 91.5 | 91.6 | 91.7 | 91.8 |
| 6.0 | 85.4 | 86.0 | 86.4 | 87.0 | 87.4 | 87.7 | 87.9 | 87.9 | 88.1 | 88.4 | 88.6 |
| 7.0 | 81.0 | 81.8 | 82.3 | 83.1 | 83.5 | 83.9 | 84.3 | 84.4 | 84.7 | 85.2 | 85.5 |
| 8.0 | 76.9 | 77.7 | 78.3 | 79.3 | 79.8 | 80.3 | 80.7 | 81.0 | 81.4 | 81.8 | 82.1 |
| 9.0 | 72.9 | 73.8 | 74.5 | 75.6 | 76.3 | 76.8 | 77.2 | 77.7 | 78.1 | 78.5 | 78.9 |
| 10.0 | 69.1 | 70.1 | 70.8 | 72.0 | 72.8 | 73.4 | 73.9 | 74.4 | 75.0 | 75.4 | 75.8 |
| 12.0 | 62.4 | 63.4 | 64.2 | 65.4 | 66.3 | 67.0 | 67.7 | 68.4 | 68.9 | 69.4 | 69.8 |
| 14.0 | 56.2 | 57.2 | 57.9 | 59.2 | 60.2 | 61.0 | 61.8 | 62.7 | 63.3 | 63.8 | 64.2 |
| 16.0 | 50.7 | 51.6 | 52.3 | 53.6 | 54.6 | 55.5 | 56.5 | 57.4 | 58.1 | 58.7 | 59.1 |
| 18.0 | 45.7 | 46.5 | 47.3 | 48.6 | 49.7 | 50.6 | 51.6 | 52.6 | 53.3 | 54.0 | 54.4 |
| 20.0 | 41.1 | 41.8 | 42.6 | 43.9 | 45.0 | 45.9 | 47.0 | 48.1 | 48.9 | 49.5 | 50.0 |
| 22.0 | 37.2 | 37.9 | 38.6 | 39.9 | 40.9 | 41.8 | 42.9 | 44.0 | 44.9 | 45.6 | 46.1 |
| 24.0 | 33.6 | 34.3 | 34.9 | 36.1 | 37.1 | 37.9 | 39.1 | 40.2 | 41.1 | 41.8 | 42.4 |
| 26.0 | 30.4 | 31.0 | 31.6 | 32.7 | 33.6 | 34.5 | 35.6 | 36.8 | 37.7 | 38.4 | 39.0 |
| 28.0 | 27.5 | 28.1 | 28.7 | 29.8 | 30.6 | 31.4 | 32.5 | 33.7 | 34.6 | 35.3 | 35.9 |
| 30.0 | 24.7 | 25.3 | 25.9 | 26.9 | 27.7 | 28.5 | 29.6 | 30.8 | 31.6 | 32.3 | 32.9 |

Adapted from The University of Iowa Hospitals and Clinics, Iowa City, IA, 1992, personal communication.

## TABLE 1.D4
## Tissue Phantom Ratios for 10-MV X Rays [Varian Clinac-18, 100 cm SAD]

| Depth (cm) | Field Size (cm×cm) | | | | | | | | | | |
|---|---|---|---|---|---|---|---|---|---|---|---|
| | 4×4 | 5×5 | 6×6 | 8×8 | 10×10 | 12×12 | 15×15 | 20×20 | 25×25 | 30×30 | 35×35 |
| 1.0 | 0.868 | 0.875 | 0.879 | 0.887 | 0.896 | 0.903 | 0.912 | 0.932 | 0.946 | 0.952 | 0.954 |
| 2.0 | 0.991 | 0.991 | 0.991 | 0.993 | 0.995 | 0.997 | 0.999 | 1.001 | 1.000 | 0.999 | 0.999 |
| 2.5 | 1.000 | 1.000 | 1.000 | 1.000 | 1.000 | 1.000 | 1.000 | 1.000 | 1.000 | 1.000 | 1.000 |
| 3.0 | 0.992 | 0.993 | 0.993 | 0.993 | 0.994 | 0.993 | 0.991 | 0.991 | 0.989 | 0.990 | 0.992 |
| 4.0 | 0.966 | 0.970 | 0.971 | 0.974 | 0.978 | 0.980 | 0.979 | 0.976 | 0.977 | 0.979 | 0.980 |
| 5.0 | 0.943 | 0.948 | 0.951 | 0.954 | 0.957 | 0.960 | 0.962 | 0.960 | 0.960 | 0.961 | 0.962 |
| 6.0 | 0.911 | 0.918 | 0.922 | 0.928 | 0.932 | 0.936 | 0.939 | 0.939 | 0.940 | 0.943 | 0.946 |
| 7.0 | 0.880 | 0.888 | 0.894 | 0.902 | 0.907 | 0.912 | 0.916 | 0.919 | 0.921 | 0.925 | 0.929 |
| 8.0 | 0.850 | 0.859 | 0.866 | 0.876 | 0.882 | 0.888 | 0.893 | 0.897 | 0.900 | 0.905 | 0.909 |
| 9.0 | 0.820 | 0.831 | 0.838 | 0.849 | 0.857 | 0.864 | 0.870 | 0.875 | 0.879 | 0.884 | 0.889 |
| 10.0 | 0.790 | 0.802 | 0.811 | 0.823 | 0.833 | 0.840 | 0.847 | 0.854 | 0.859 | 0.864 | 0.868 |
| 12.0 | 0.739 | 0.751 | 0.760 | 0.773 | 0.783 | 0.792 | 0.801 | 0.811 | 0.817 | 0.823 | 0.828 |
| 14.0 | 0.689 | 0.700 | 0.709 | 0.723 | 0.734 | 0.745 | 0.756 | 0.769 | 0.776 | 0.782 | 0.788 |
| 16.0 | 0.642 | 0.652 | 0.661 | 0.675 | 0.688 | 0.699 | 0.712 | 0.727 | 0.736 | 0.743 | 0.750 |
| 18.0 | 0.599 | 0.608 | 0.617 | 0.631 | 0.644 | 0.657 | 0.670 | 0.687 | 0.697 | 0.705 | 0.712 |
| 20.0 | 0.556 | 0.564 | 0.572 | 0.587 | 0.601 | 0.614 | 0.628 | 0.647 | 0.658 | 0.667 | 0.675 |
| 22.0 | 0.520 | 0.528 | 0.536 | 0.549 | 0.563 | 0.576 | 0.590 | 0.610 | 0.622 | 0.631 | 0.640 |
| 24.0 | 0.484 | 0.492 | 0.499 | 0.512 | 0.525 | 0.537 | 0.552 | 0.573 | 0.586 | 0.596 | 0.606 |
| 26.0 | 0.450 | 0.458 | 0.465 | 0.478 | 0.490 | 0.502 | 0.517 | 0.539 | 0.552 | 0.563 | 0.573 |
| 28.0 | 0.419 | 0.427 | 0.434 | 0.447 | 0.459 | 0.470 | 0.484 | 0.506 | 0.521 | 0.532 | 0.542 |
| 30.0 | 0.388 | 0.395 | 0.402 | 0.416 | 0.427 | 0.438 | 0.452 | 0.474 | 0.490 | 0.501 | 0.510 |

Adapted from The University of Iowa Hospitals and Clinics, Iowa City, IA, 1992, personal communication.

**TABLE 1.D5**

**Percent Depth Doses for 10-MV X Rays [Siemens Mevatron XII, 100 cm SSD]**

| Depth (cm) | Field Size (cm×cm) | | | | | | | | | |
|---|---|---|---|---|---|---|---|---|---|---|
| | 0×0 | 4×4 | 5×5 | 6×6 | 8×8 | 10×10 | 12×12 | 15×15 | 20×20 | 30×30 |
| 0 | 3.0 | 6.0 | 7.4 | 8.5 | 10.2 | 11.8 | 12.7 | 14.1 | 15.9 | 18.3 |
| 1.0 | 69.2 | 75.5 | 76.8 | 78.0 | 80.2 | 82.0 | 83.6 | 85.6 | 88.0 | 90.8 |
| 1.5 | 92.4 | 92.9 | 93.1 | 93.2 | 93.5 | 93.7 | 94.0 | 94.4 | 95.0 | 96.3 |
| 2.5 | 100.0 | 100.0 | 100.0 | 100.0 | 100.0 | 100.0 | 100.0 | 100.0 | 100.0 | 100.0 |
| 3.0 | 98.6 | 99.1 | 99.2 | 99.3 | 99.4 | 99.5 | 99.6 | 99.7 | 99.8 | 99.9 |
| 4.0 | 92.1 | 95.1 | 95.5 | 95.9 | 96.4 | 96.8 | 97.0 | 97.2 | 97.4 | 97.5 |
| 5.0 | 87.0 | 90.6 | 91.1 | 91.6 | 92.4 | 93.0 | 93.4 | 93.8 | 94.2 | 94.4 |
| 6.0 | 82.9 | 86.2 | 86.8 | 87.3 | 88.2 | 88.9 | 89.4 | 90.0 | 90.7 | 91.2 |
| 7.0 | 78.7 | 82.1 | 82.7 | 83.3 | 84.2 | 85.0 | 85.6 | 86.3 | 87.1 | 87.8 |
| 8.0 | 74.7 | 78.2 | 78.8 | 79.4 | 80.5 | 81.3 | 82.0 | 82.8 | 83.6 | 84.5 |
| 9.0 | 70.9 | 74.4 | 75.1 | 75.8 | 76.8 | 77.7 | 78.4 | 79.3 | 80.3 | 81.2 |
| 10.0 | 67.2 | 70.9 | 71.6 | 72.2 | 73.4 | 74.3 | 75.1 | 76.0 | 77.0 | 78.1 |
| 11.0 | 65.3 | 68.0 | 68.6 | 69.1 | 70.1 | 70.9 | 71.7 | 72.6 | 73.8 | 75.2 |
| 12.0 | 60.5 | 64.3 | 65.0 | 65.7 | 66.9 | 67.9 | 68.8 | 69.8 | 70.9 | 72.2 |
| 13.0 | 57.4 | 61.2 | 62.0 | 62.7 | 63.9 | 65.0 | 65.8 | 66.9 | 68.1 | 69.4 |
| 14.0 | 54.4 | 58.3 | 59.1 | 59.8 | 61.0 | 62.1 | 63.0 | 64.0 | 65.3 | 66.7 |
| 15.0 | 51.4 | 55.4 | 56.2 | 57.0 | 58.3 | 59.4 | 60.3 | 61.4 | 62.7 | 64.1 |
| 16.0 | 48.8 | 52.8 | 53.6 | 54.4 | 55.7 | 56.8 | 57.7 | 58.8 | 60.2 | 61.6 |
| 17.0 | 46.3 | 50.3 | 51.1 | 51.8 | 53.1 | 54.3 | 55.2 | 56.3 | 57.7 | 59.2 |
| 18.0 | 44.1 | 48.0 | 48.7 | 49.5 | 50.8 | 51.9 | 52.8 | 54.0 | 55.4 | 57.0 |
| 19.0 | 41.5 | 45.6 | 46.4 | 47.1 | 48.5 | 49.6 | 50.6 | 51.8 | 53.2 | 54.8 |
| 20.0 | 39.4 | 43.4 | 44.2 | 45.0 | 46.3 | 47.5 | 48.4 | 49.6 | 51.0 | 52.6 |
| 21.0 | 37.3 | 41.3 | 42.1 | 42.9 | 44.2 | 45.3 | 46.3 | 47.5 | 49.0 | 50.6 |
| 22.0 | 35.4 | 39.4 | 40.2 | 40.9 | 42.2 | 43.4 | 44.3 | 45.5 | 47.0 | 48.6 |
| 23.0 | 33.5 | 37.5 | 38.3 | 39.0 | 40.3 | 41.5 | 42.4 | 43.6 | 45.1 | 46.7 |
| 24.0 | 31.8 | 35.7 | 36.5 | 37.2 | 38.5 | 39.6 | 40.6 | 41.8 | 43.3 | 45.0 |
| 26.0 | 28.6 | 32.4 | 33.1 | 33.9 | 35.1 | 36.3 | 37.2 | 38.4 | 39.9 | 41.6 |
| 28.0 | 25.6 | 29.3 | 30.1 | 30.8 | 32.1 | 33.2 | 34.1 | 35.3 | 36.7 | 38.4 |
| 30.0 | 23.1 | 26.6 | 27.4 | 28.0 | 29.3 | 30.3 | 31.2 | 32.4 | 33.8 | 35.4 |

From Keller, B., Bassano, D., Mathewson, C., and Rubin, P., 10 MV photon beam characteristics: central axis depth doses, tissue-maximum ratios, scatter-maximum ratios, beam: flatness, backscatter, and output factors, *Int. J. Radiat. Oncol. Biol. Phys.*, 1, 69, 1975. With permission.

## TABLE 1.D6
## Tissue Phantom Ratios for 10-MV X Rays [Siemens Mevatron XII, 100 cm SAD]

| Depth (cm) | Field Size (cm × cm) | | | | | | | | | |
|---|---|---|---|---|---|---|---|---|---|---|
| | 0×0 | 4×4 | 5×5 | 6×6 | 8×8 | 10×10 | 12×12 | 15×15 | 20×20 | 30×30 |
| 1.0 | 0.672 | 0.733 | 0.745 | 0.757 | 0.778 | 0.796 | 0.811 | 0.830 | 0.854 | 0.881 |
| 1.5 | 0.906 | 0.911 | 0.913 | 0.914 | 0.916 | 0.919 | 0.922 | 0.925 | 0.932 | 0.944 |
| 2.5 | 1.000 | 1.000 | 1.000 | 1.000 | 1.000 | 1.000 | 1.000 | 1.000 | 1.000 | 1.000 |
| 3.0 | 0.996 | 1.001 | 1.002 | 1.003 | 1.004 | 1.005 | 1.006 | 1.007 | 1.008 | 1.008 |
| 4.0 | 0.948 | 0.978 | 0.982 | 0.986 | 0.992 | 0.996 | 0.998 | 1.001 | 1.003 | 1.004 |
| 5.0 | 0.913 | 0.949 | 0.955 | 0.960 | 0.968 | 0.974 | 0.979 | 0.984 | 0.988 | 0.991 |
| 6.0 | 0.887 | 0.920 | 0.926 | 0.932 | 0.941 | 0.949 | 0.955 | 0.961 | 0.968 | 0.975 |
| 7.0 | 0.858 | 0.892 | 0.899 | 0.905 | 0.915 | 0.924 | 0.930 | 0.938 | 0.947 | 0.955 |
| 8.0 | 0.829 | 0.865 | 0.872 | 0.879 | 0.890 | 0.899 | 0.907 | 0.916 | 0.926 | 0.936 |
| 9.0 | 0.802 | 0.839 | 0.846 | 0.853 | 0.865 | 0.875 | 0.883 | 0.893 | 0.904 | 0.917 |
| 10.0 | 0.774 | 0.813 | 0.820 | 0.828 | 0.840 | 0.851 | 0.859 | 0.870 | 0.883 | 0.896 |
| 11.0 | 0.766 | 0.795 | 0.801 | 0.807 | 0.817 | 0.826 | 0.835 | 0.846 | 0.860 | 0.878 |
| 12.0 | 0.722 | 0.763 | 0.771 | 0.779 | 0.792 | 0.804 | 0.814 | 0.826 | 0.841 | 0.858 |
| 13.0 | 0.698 | 0.739 | 0.747 | 0.755 | 0.769 | 0.781 | 0.792 | 0.805 | 0.820 | 0.839 |
| 14.0 | 0.673 | 0.715 | 0.724 | 0.732 | 0.746 | 0.759 | 0.770 | 0.783 | 0.800 | 0.820 |
| 15.0 | 0.647 | 0.691 | 0.700 | 0.709 | 0.724 | 0.737 | 0.748 | 0.763 | 0.780 | 0.801 |
| 16.0 | 0.625 | 0.669 | 0.678 | 0.687 | 0.702 | 0.716 | 0.727 | 0.742 | 0.760 | 0.782 |
| 17.0 | 0.603 | 0.648 | 0.657 | 0.665 | 0.681 | 0.695 | 0.707 | 0.722 | 0.741 | 0.764 |
| 18.0 | 0.584 | 0.627 | 0.637 | 0.645 | 0.661 | 0.674 | 0.686 | 0.701 | 0.721 | 0.746 |
| 19.0 | 0.559 | 0.605 | 0.615 | 0.624 | 0.640 | 0.654 | 0.667 | 0.683 | 0.703 | 0.728 |
| 20.0 | 0.540 | 0.585 | 0.595 | 0.604 | 0.620 | 0.635 | 0.647 | 0.664 | 0.684 | 0.710 |
| 22.0 | 0.502 | 0.547 | 0.556 | 0.565 | 0.582 | 0.597 | 0.610 | 0.626 | 0.648 | 0.676 |
| 24.0 | 0.465 | 0.510 | 0.520 | 0.529 | 0.545 | 0.560 | 0.573 | 0.590 | 0.613 | 0.642 |
| 25.0 | 0.446 | 0.492 | 0.501 | 0.511 | 0.528 | 0.545 | 0.556 | 0.573 | 0.596 | 0.625 |
| 26.0 | 0.432 | 0.476 | 0.485 | 0.494 | 0.511 | 0.525 | 0.539 | 0.556 | 0.579 | 0.610 |
| 28.0 | 0.399 | 0.443 | 0.452 | 0.461 | 0.477 | 0.492 | 0.506 | 0.523 | 0.546 | 0.578 |
| 30.0 | 0.372 | 0.413 | 0.422 | 0.431 | 0.447 | 0.461 | 0.474 | 0.491 | 0.515 | 0.547 |

From Keller, B., Bassano, D., Mathewson, C., and Rubin, P., 10 MV photon beam characteristics: central axis depth doses, tissue-maximum ratios, scatter-maximum ratios, beam flatness, backscatter, and output factors, *Int. J. Radiat. Oncol. Biol. Phys.*, 1, 69, 1975. With permission.

**TABLE 1.D7**

**Percent Depth Doses for 10-MV X Rays [Toshiba LMR-13, 100 cm SSD]**

| Depth (cm) | Field Size (cm×cm) | | | | | | | | | |
|---|---|---|---|---|---|---|---|---|---|---|
| | 0×0 | 4×4 | 6×6 | 8×8 | 10×10 | 12×12 | 15×15 | 20×20 | 25×25 | 30×30 |
| 0.0 | 5.0 | 6.5 | 8.5 | 10.7 | 12.5 | 14.5 | 17.0 | 21.0 | 24.5 | 28.0 |
| 0.2 | 37.0 | 40.0 | 43.0 | 45.0 | 46.5 | 48.0 | 50.0 | 52.5 | 54.0 | 56.0 |
| 0.5 | 65.0 | 67.0 | 69.0 | 70.5 | 72.0 | 73.0 | 74.0 | 76.0 | 77.0 | 79.0 |
| 1.0 | 86.0 | 88.0 | 89.0 | 90.0 | 91.0 | 91.5 | 92.0 | 93.0 | 94.0 | 95.0 |
| 1.5 | 94.5 | 95.5 | 96.0 | 96.5 | 97.0 | 97.0 | 97.5 | 98.0 | 98.0 | 98.5 |
| 2.0 | 96.5 | 97.5 | 98.0 | 98.0 | 98.0 | 98.5 | 99.0 | 99.0 | 99.5 | 99.5 |
| 2.5 | 100.0 | 100.0 | 100.0 | 100.0 | 100.0 | 100.0 | 100.0 | 100.0 | 100.0 | 100.0 |
| 3.0 | 97.4 | 99.0 | 99.0 | 99.0 | 99.0 | 99.0 | 99.0 | 99.0 | 99.0 | 99.0 |
| 4.0 | 92.3 | 96.4 | 96.4 | 96.4 | 96.4 | 96.5 | 96.5 | 96.5 | 96.5 | 96.5 |
| 5.0 | 87.5 | 91.6 | 91.8 | 91.9 | 92.1 | 92.2 | 92.3 | 92.5 | 92.6 | 92.7 |
| 6.0 | 83.0 | 87.0 | 87.4 | 87.7 | 87.9 | 88.1 | 88.3 | 88.6 | 88.8 | 89.0 |
| 7.0 | 78.7 | 82.6 | 83.2 | 83.6 | 83.9 | 84.2 | 84.5 | 84.9 | 85.2 | 85.5 |
| 8.0 | 74.7 | 78.5 | 79.2 | 79.7 | 80.1 | 80.4 | 80.8 | 81.4 | 81.8 | 82.1 |
| 9.0 | 70.8 | 74.6 | 75.4 | 76.0 | 76.5 | 76.9 | 77.3 | 78.0 | 78.4 | 78.8 |
| 10.0 | 67.2 | 70.8 | 71.8 | 72.5 | 73.0 | 73.5 | 74.0 | 74.7 | 75.3 | 75.7 |
| 11.0 | 63.8 | 67.3 | 68.4 | 69.1 | 69.7 | 70.2 | 70.8 | 71.6 | 72.2 | 72.7 |
| 12.0 | 60.6 | 63.9 | 65.1 | 65.9 | 66.6 | 67.1 | 67.7 | 68.6 | 69.3 | 69.8 |
| 13.0 | 57.5 | 60.7 | 62.0 | 62.8 | 63.5 | 64.1 | 64.8 | 65.7 | 66.5 | 67.1 |
| 14.0 | 54.6 | 57.7 | 59.0 | 59.9 | 60.7 | 61.3 | 62.0 | 63.0 | 63.8 | 64.4 |
| 15.0 | 51.9 | 54.8 | 56.2 | 57.1 | 57.9 | 58.5 | 59.3 | 60.4 | 61.2 | 61.8 |
| 16.0 | 49.3 | 52.1 | 53.5 | 54.5 | 55.3 | 55.9 | 56.8 | 57.8 | 58.7 | 59.4 |
| 17.0 | 46.8 | 49.5 | 50.9 | 52.0 | 52.8 | 53.5 | 54.3 | 55.4 | 56.3 | 57.0 |
| 18.0 | 44.5 | 47.0 | 48.5 | 49.5 | 50.4 | 51.1 | 52.0 | 53.1 | 54.0 | 54.8 |
| 19.0 | 42.3 | 44.7 | 46.1 | 47.2 | 48.1 | 48.8 | 49.7 | 50.9 | 51.8 | 52.6 |
| 20.0 | 40.2 | 42.4 | 43.9 | 45.0 | 45.9 | 46.7 | 47.6 | 48.8 | 49.7 | 50.5 |
| 22.0 | 36.3 | 38.3 | 39.8 | 41.0 | 41.9 | 42.6 | 43.5 | 44.8 | 45.8 | 46.6 |
| 24.0 | 32.8 | 34.6 | 36.1 | 37.2 | 38.2 | 38.9 | 39.9 | 41.1 | 42.1 | 43.0 |
| 26.0 | 29.7 | 31.2 | 32.7 | 33.9 | 34.8 | 35.5 | 36.5 | 37.8 | 38.8 | 39.6 |
| 28.0 | 26.9 | 28.1 | 29.7 | 30.8 | 31.7 | 32.5 | 33.4 | 34.7 | 35.7 | 36.5 |
| 30.0 | 24.3 | 25.4 | 26.9 | 28.0 | 28.9 | 29.6 | 30.6 | 31.8 | 32.9 | 33.7 |

From Khan, F.M., Depth dose and scatter analysis of 10 MV x-rays, *Radiol.*, 106, 662, 1973. With permission.

## TABLE 1.D8
### Tissue Phantom Ratios for 10-MV X Rays [Toshiba LMR-13, 100 cm SAD]

| Depth (cm) | Field Size (cm×cm) | | | | | | | | | |
|---|---|---|---|---|---|---|---|---|---|---|
| | 0×0 | 4×4 | 6×6 | 8×8 | 10×10 | 12×12 | 15×15 | 20×20 | 25×25 | 30×30 |
| 0.0 | 0.048 | 0.062 | 0.081 | 0.102 | 0.119 | 0.138 | 0.162 | 0.200 | 0.233 | 0.267 |
| 0.2 | 0.354 | 0.382 | 0.411 | 0.430 | 0.444 | 0.459 | 0.478 | 0.502 | 0.516 | 0.535 |
| 0.5 | 0.625 | 0.644 | 0.663 | 0.678 | 0.692 | 0.702 | 0.711 | 0.731 | 0.740 | 0.759 |
| 1.0 | 0.835 | 0.854 | 0.864 | 0.874 | 0.884 | 0.888 | 0.893 | 0.903 | 0.913 | 0.922 |
| 1.5 | 0.927 | 0.936 | 0.941 | 0.946 | 0.951 | 0.951 | 0.956 | 0.961 | 0.961 | 0.966 |
| 2.0 | 0.956 | 0.966 | 0.970 | 0.970 | 0.970 | 0.975 | 0.980 | 0.980 | 0.985 | 0.985 |
| 2.5 | 1.000 | 1.000 | 1.000 | 1.000 | 1.000 | 1.000 | 1.000 | 1.000 | 1.000 | 1.000 |
| 3.0 | 0.983 | 1.000 | 1.000 | 1.000 | 1.000 | 1.000 | 1.000 | 1.000 | 1.000 | 1.000 |
| 4.0 | 0.950 | 0.992 | 0.992 | 0.993 | 0.993 | 0.993 | 0.993 | 0.993 | 0.993 | 0.994 |
| 5.0 | 0.918 | 0.960 | 0.963 | 0.965 | 0.966 | 0.967 | 0.968 | 0.970 | 0.971 | 0.972 |
| 6.0 | 0.887 | 0.930 | 0.934 | 0.937 | 0.939 | 0.941 | 0.944 | 0.947 | 0.949 | 0.951 |
| 7.0 | 0.858 | 0.899 | 0.906 | 0.910 | 0.913 | 0.916 | 0.920 | 0.924 | 0.928 | 0.931 |
| 8.0 | 0.829 | 0.870 | 0.878 | 0.884 | 0.888 | 0.892 | 0.896 | 0.902 | 0.906 | 0.910 |
| 9.0 | 0.801 | 0.841 | 0.851 | 0.858 | 0.863 | 0.867 | 0.873 | 0.880 | 0.885 | 0.889 |
| 10.0 | 0.774 | 0.813 | 0.824 | 0.832 | 0.838 | 0.843 | 0.850 | 0.858 | 0.864 | 0.869 |
| 11.0 | 0.748 | 0.786 | 0.798 | 0.807 | 0.814 | 0.820 | 0.827 | 0.836 | 0.843 | 0.849 |
| 12.0 | 0.723 | 0.760 | 0.773 | 0.783 | 0.791 | 0.797 | 0.805 | 0.815 | 0.823 | 0.830 |
| 13.0 | 0.699 | 0.734 | 0.749 | 0.759 | 0.768 | 0.774 | 0.783 | 0.794 | 0.803 | 0.810 |
| 14.0 | 0.676 | 0.709 | 0.725 | 0.736 | 0.745 | 0.752 | 0.762 | 0.774 | 0.783 | 0.791 |
| 15.0 | 0.653 | 0.684 | 0.701 | 0.713 | 0.723 | 0.731 | 0.741 | 0.753 | 0.764 | 0.772 |
| 16.0 | 0.631 | 0.661 | 0.678 | 0.691 | 0.701 | 0.710 | 0.720 | 0.734 | 0.744 | 0.753 |
| 17.0 | 0.610 | 0.638 | 0.656 | 0.669 | 0.680 | 0.689 | 0.700 | 0.714 | 0.726 | 0.735 |
| 18.0 | 0.589 | 0.615 | 0.634 | 0.648 | 0.659 | 0.669 | 0.680 | 0.695 | 0.707 | 0.717 |
| 19.0 | 0.570 | 0.593 | 0.613 | 0.628 | 0.639 | 0.649 | 0.661 | 0.676 | 0.689 | 0.699 |
| 20.0 | 0.551 | 0.572 | 0.593 | 0.608 | 0.620 | 0.629 | 0.642 | 0.658 | 0.671 | 0.681 |
| 22.0 | 0.514 | 0.532 | 0.553 | 0.569 | 0.582 | 0.592 | 0.605 | 0.622 | 0.636 | 0.647 |
| 24.0 | 0.480 | 0.494 | 0.516 | 0.533 | 0.546 | 0.556 | 0.570 | 0.588 | 0.602 | 0.614 |
| 26.0 | 0.449 | 0.458 | 0.481 | 0.498 | 0.511 | 0.522 | 0.536 | 0.555 | 0.570 | 0.583 |
| 28.0 | 0.419 | 0.425 | 0.448 | 0.465 | 0.479 | 0.490 | 0.505 | 0.524 | 0.539 | 0.552 |
| 30.0 | 0.392 | 0.394 | 0.417 | 0.434 | 0.448 | 0.459 | 0.474 | 0.494 | 0.509 | 0.523 |

From Khan, F.M., Depth dose and scatter analysis of 10 MV x-rays, *Radiol.*, 106, 662, 1973. With permission.

**TABLE 1.E1**

**Percent Depth Doses for 14-MV X Rays [Toshiba LMR-16, 100 cm SSD]**

| Depth (cm) | Field Size (cm×cm) | | | | | | | | | |
|---|---|---|---|---|---|---|---|---|---|---|
| | 5×5 | 8×8 | 10×10 | 12×12 | 15×15 | 18×18 | 20×20 | 25×25 | 28×28 |
| 0.01 | 36.0 | 38.1 | 40.2 | 42.2 | 45.0 | 47.5 | 49.6 | 52.0 | 54.3 |
| 0.50 | 71.5 | 71.6 | 72.4 | 75.0 | 76.6 | 78.2 | 80.5 | 82.5 | 82.2 |
| 1.00 | 88.0 | 88.2 | 88.9 | 90.0 | 91.0 | 91.9 | 92.6 | 94.0 | 94.3 |
| 1.50 | 94.5 | 95.4 | 95.8 | 96.7 | 97.0 | 97.4 | 98.0 | 99.2 | 98.1 |
| 2.00 | 98.8 | 99.4 | 99.3 | 99.2 | 100.0 | 99.4 | 99.9 | 99.7 | 100.0 |
| 2.50 | 100.0 | 99.8 | 100.0 | 99.5 | 100.0 | 100.0 | 99.8 | 99.7 | 99.5 |
| 3.00 | 100.0 | 99.5 | 99.5 | 99.7 | 98.8 | 99.0 | 99.3 | 98.6 | 99.5 |
| 3.50 | 99.0 | 99.1 | 98.7 | 98.2 | 97.5 | 98.2 | 98.4 | 97.8 | 97.6 |
| 4.00 | 97.5 | 97.2 | 97.8 | 97.2 | 96.3 | 96.6 | 96.4 | 96.0 | 96.0 |
| 5.00 | 93.1 | 93.6 | 93.9 | 93.4 | 93.3 | 93.0 | 93.2 | 92.6 | 93.2 |
| 10.00 | 73.1 | 75.1 | 76.0 | 76.1 | 76.3 | 76.8 | 77.2 | 77.1 | 77.7 |
| 15.00 | 58.1 | 60.5 | 61.2 | 62.2 | 62.2 | 62.9 | 63.2 | 63.8 | 64.3 |
| 20.00 | 46.3 | 48.1 | 49.2 | 50.1 | 50.8 | 51.9 | 51.8 | 52.9 | 53.3 |
| 25.00 | 37.2 | 38.9 | 40.1 | 40.7 | 41.8 | 42.6 | 42.5 | 43.7 | 44.8 |
| 30.00 | 30.1 | 31.6 | 32.5 | 33.4 | 34.2 | 35.1 | 35.2 | 36.0 | 36.3 |
| 35.00 | 24.8 | 25.9 | 26.6 | 27.0 | 28.1 | 28.8 | 29.0 | 29.3 | 29.9 |

From Mantel, J., Perry, H., and Weinkam, J.J., X-ray depth-dose characteristics of the Toshiba LMR-16, *Med. Phys.*, 6, 95, 1979. With permission.

## TABLE 1.E2
### Tissue Phantom Ratios for 14-MV X Rays [Toshiba LMR-16, 100 cm SAD]

| Depth (cm) | Field Size (cm×cm) | | | | | | | | | |
|---|---|---|---|---|---|---|---|---|---|---|
| | 5×5 | 8×8 | 10×10 | 12×12 | 15×15 | 18×18 | 20×20 | 25×25 | 28×28 |
| 0.01 | 0.346 | 0.370 | 0.386 | 0.403 | 0.428 | 0.453 | 0.470 | 0.512 | 0.537 |
| 0.50 | 0.668 | 0.680 | 0.692 | 0.705 | 0.724 | 0.744 | 0.756 | 0.785 | 0.801 |
| 1.00 | 0.845 | 0.850 | 0.858 | 0.867 | 0.880 | 0.893 | 0.901 | 0.917 | 0.925 |
| 1.50 | 0.935 | 0.938 | 0.943 | 0.949 | 0.957 | 0.965 | 0.969 | 0.977 | 0.980 |
| 2.00 | 0.979 | 0.981 | 0.984 | 0.988 | 0.993 | 0.997 | 0.999 | 1.001 | 1.002 |
| 2.50 | 0.997 | 0.999 | 1.002 | 1.004 | 1.007 | 1.008 | 1.009 | 1.008 | 1.007 |
| 3.00 | 1.000 | 1.003 | 1.005 | 1.007 | 1.009 | 1.009 | 1.009 | 1.007 | 1.005 |
| 3.50 | 0.996 | 1.000 | 1.002 | 1.004 | 1.004 | 1.004 | 1.004 | 1.001 | 0.999 |
| 4.00 | 0.987 | 0.993 | 0.995 | 0.996 | 0.997 | 0.997 | 0.996 | 0.994 | 0.991 |
| 5.00 | 0.965 | 0.972 | 0.975 | 0.977 | 0.978 | 0.979 | 0.978 | 0.976 | 0.974 |
| 10.00 | 0.839 | 0.856 | 0.863 | 0.869 | 0.875 | 0.879 | 0.881 | 0.884 | 0.885 |
| 15.00 | 0.725 | 0.749 | 0.760 | 0.768 | 0.778 | 0.785 | 0.789 | 0.797 | 0.800 |
| 20.00 | 0.624 | 0.652 | 0.666 | 0.676 | 0.689 | 0.699 | 0.704 | 0.715 | 0.721 |
| 25.00 | 0.535 | 0.566 | 0.581 | 0.593 | 0.608 | 0.620 | 0.626 | 0.640 | 0.646 |
| 30.00 | 0.458 | 0.490 | 0.506 | 0.519 | 0.534 | 0.547 | 0.555 | 0.570 | 0.578 |
| 35.00 | 0.390 | 0.422 | 0.438 | 0.452 | 0.468 | 0.482 | 0.490 | 0.506 | 0.513 |

From Mantel, J., Perry, H., and Weinkam, J.J., X-ray depth-dose characteristics of the Toshiba LMR-16, *Med. Phys.*, 6, 95, 1979. With permission.

## TABLE 1.E3
### Tissue Phantom Ratios for 15-MV X Rays [Mevatron-77, 100 cm SAD]

| Depth (cm) | Field Size (cm×cm) | | | | | | | | |
|---|---|---|---|---|---|---|---|---|---|
| | 0×0 | 4×4 | 6×6 | 10×10 | 15×15 | 20×20 | 25×25 | 30×30 | 35×35 |
| 0.0 | 0.050 | 0.055 | 0.070 | 0.100 | 0.155 | 0.210 | 0.245 | 0.275 | 0.290 |
| 0.5 | 0.525 | 0.530 | 0.540 | 0.565 | 0.600 | 0.640 | 0.675 | 0.690 | 0.700 |
| 1.0 | 0.720 | 0.720 | 0.725 | 0.740 | 0.780 | 0.805 | 0.825 | 0.840 | 0.845 |
| 1.5 | 0.870 | 0.870 | 0.870 | 0.880 | 0.890 | 0.910 | 0.925 | 0.935 | 0.940 |
| 2.0 | 0.950 | 0.950 | 0.950 | 0.950 | 0.955 | 0.965 | 0.975 | 0.980 | 0.980 |
| 2.5 | 0.975 | 0.975 | 0.975 | 0.980 | 0.985 | 0.990 | 0.995 | 0.995 | 0.995 |
| 2.75 | 0.980 | 0.980 | 0.985 | 0.990 | 0.995 | 1.000 | 1.000 | 1.000 | 1.000 |
| 3.0 | 1.000 | 1.000 | 1.000 | 1.000 | 1.000 | 1.000 | 1.000 | 1.000 | 1.000 |
| 4.0 | 0.970 | 0.990 | 0.990 | 0.990 | 0.990 | 0.990 | 0.990 | 0.990 | 0.990 |
| 5.0 | 0.940 | 0.975 | 0.980 | 0.980 | 0.980 | 0.980 | 0.980 | 0.980 | 0.980 |
| 6.0 | 0.915 | 0.950 | 0.960 | 0.965 | 0.965 | 0.965 | 0.965 | 0.965 | 0.965 |
| 8.0 | 0.860 | 0.900 | 0.910 | 0.920 | 0.925 | 0.930 | 0.930 | 0.930 | 0.935 |
| 10.0 | 0.810 | 0.850 | 0.860 | 0.875 | 0.880 | 0.890 | 0.895 | 0.895 | 0.900 |
| 15.0 | 0.700 | 0.735 | 0.750 | 0.775 | 0.790 | 0.800 | 0.810 | 0.820 | 0.830 |
| 20.0 | 0.600 | 0.635 | 0.650 | 0.675 | 0.695 | 0.705 | 0.715 | 0.725 | 0.735 |
| 25.0 | 0.515 | 0.550 | 0.565 | 0.590 | 0.610 | 0.625 | 0.635 | 0.645 | 0.655 |
| 30.0 | 0.445 | 0.475 | 0.490 | 0.515 | 0.535 | 0.550 | 0.565 | 0.575 | 0.580 |

From Paul, J.M., Koch, R.F., Khan, F.R., and Devi, B.S., Characteristics of Mevatron 77 15-MV photon beam, *Med. Phys.*, 10, 237, 1983. With permission.

## TABLE 1.E4
### Percent Depth Doses for 18-MV X Rays [Varian Clinac-1800, 100 cm SSD]

| Depth (cm) | Field Size (cm×cm) | | | | | | | | | | |
| --- | --- | --- | --- | --- | --- | --- | --- | --- | --- | --- | --- |
| | 4×4 | 5×5 | 6×6 | 8×8 | 10×10 | 12×12 | 15×15 | 20×20 | 25×25 | 30×30 | 35×35 |
| 3.2 | 99.3 | 99.1 | 99.5 | 99.8 | 99.9 | 100.0 | 100.0 | 99.8 | 99.7 | 99.5 | 99.3 |
| 3.6 | 100.0 | 100.0 | 100.0 | 100.0 | 100.0 | 99.7 | 99.6 | 99.1 | 98.8 | 98.4 | 98.6 |
| 4.0 | 99.6 | 99.7 | 99.8 | 99.7 | 99.7 | 99.0 | 99.0 | 98.2 | 97.6 | 97.2 | 97.5 |
| 5.0 | 97.5 | 97.7 | 97.8 | 97.5 | 97.0 | 96.5 | 96.2 | 95.0 | 94.4 | 94.2 | 94.5 |
| 6.0 | 94.5 | 94.5 | 94.5 | 94.4 | 94.1 | 93.2 | 92.6 | 91.6 | 91.2 | 91.0 | 91.0 |
| 7.0 | 90.2 | 90.7 | 91.0 | 91.0 | 90.5 | 90.0 | 89.4 | 88.4 | 88.0 | 87.8 | 88.2 |
| 8.0 | 87.0 | 87.1 | 87.1 | 87.5 | 87.0 | 86.5 | 86.0 | 85.0 | 84.7 | 84.8 | 85.2 |
| 9.0 | 83.2 | 83.4 | 83.8 | 84.0 | 83.9 | 83.2 | 83.0 | 82.0 | 81.7 | 82.0 | 82.1 |
| 10.0 | 79.8 | 80.0 | 80.1 | 80.5 | 80.5 | 79.9 | 79.5 | 79.0 | 78.8 | 78.8 | 79.2 |
| 11.0 | 76.4 | 76.7 | 76.7 | 77.1 | 77.2 | 77.0 | 76.5 | 76.0 | 76.0 | 76.0 | 76.5 |
| 12.0 | 73.3 | 73.5 | 73.5 | 74.0 | 74.3 | 74.0 | 73.5 | 73.3 | 73.4 | 73.4 | 73.5 |
| 13.0 | 70.2 | 70.4 | 70.5 | 71.0 | 71.4 | 71.0 | 70.8 | 70.5 | 70.7 | 71.0 | 71.1 |
| 14.0 | 67.0 | 67.4 | 67.5 | 68.1 | 68.3 | 68.2 | 68.0 | 68.0 | 67.9 | 68.2 | 68.6 |
| 15.0 | 64.1 | 64.5 | 64.8 | 65.0 | 65.7 | 65.7 | 65.5 | 65.5 | 65.4 | 65.7 | 66.1 |
| 16.0 | 61.3 | 61.7 | 61.8 | 62.7 | 63.2 | 63.2 | 63.0 | 63.0 | 63.1 | 63.4 | 63.6 |
| 17.0 | 58.6 | 59.1 | 59.3 | 60.0 | 60.0 | 60.7 | 60.7 | 60.6 | 60.5 | 61.0 | 61.4 |
| 18.0 | 56.0 | 56.5 | 57.0 | 57.5 | 58.0 | 58.2 | 58.2 | 58.4 | 58.5 | 58.9 | 59.1 |
| 19.0 | 53.8 | 54.2 | 54.6 | 55.3 | 55.7 | 56.0 | 56.1 | 56.3 | 56.5 | 56.8 | 57.0 |
| 20.0 | 51.5 | 51.8 | 52.2 | 53.0 | 53.5 | 53.4 | 54.0 | 54.0 | 54.3 | 54.6 | 55.0 |
| 21.0 | 49.3 | 49.8 | 50.0 | 50.6 | 51.0 | 51.8 | 51.8 | 52.0 | 52.2 | 52.6 | 53.6 |
| 22.0 | 47.5 | 47.6 | 48.0 | 48.7 | 49.3 | 49.8 | 49.7 | 50.0 | 50.2 | 50.6 | 51.0 |
| 23.0 | 45.2 | 45.6 | 46.0 | 46.7 | 47.1 | 47.8 | 48.0 | 48.1 | 48.7 | 48.8 | 49.0 |
| 24.0 | 43.5 | 43.7 | 44.0 | 44.8 | 45.4 | 46.0 | 46.0 | 46.5 | 46.9 | 47.3 | 47.4 |
| 25.0 | 41.7 | 42.1 | 42.2 | 43.0 | 43.5 | 44.1 | 44.3 | 44.8 | 45.1 | 45.5 | 45.8 |
| 26.0 | 39.9 | 40.5 | 40.5 | 41.2 | 41.8 | 42.3 | 42.6 | 43.0 | 43.5 | 43.6 | 44.1 |
| 27.0 | 38.3 | 38.8 | 39.0 | 39.5 | 40.2 | 41.0 | 41.0 | 41.5 | 41.7 | 42.2 | 42.5 |
| 28.0 | 36.6 | 37.0 | 37.2 | 38.0 | 38.6 | 39.2 | 39.3 | 39.9 | 40.2 | 40.7 | 41.0 |
| 29.0 | 35.2 | 35.5 | 35.6 | 36.5 | 37.0 | 37.8 | 38.0 | 38.5 | 38.8 | 39.2 | 39.5 |
| 30.0 | 33.6 | 34.0 | 34.2 | 35.0 | 35.5 | 36.4 | 36.5 | 37.0 | 37.4 | 38.0 | 38.0 |

From Johnson, D.A., Ikoro, N.C., Chang, C.-H., Scarbrough, C.-H., Scarbrough, C.-H., Scarbrough, Ξ.C., and Antich, P.P., Properties of the 18-MV photon beam from a dual energy linear accelerator, *Med. Phys.*, 14, 1071, 1987. With permission.

## TABLE 1.E5
### Tissue Phantom Ratios for 18-MV X Rays [Varian Clinac-1800, 100 cm SAD]

| Depth (cm) | Field Size (cm×cm) | | | | | | | | | | | |
|---|---|---|---|---|---|---|---|---|---|---|---|---|
| | 0×0 | 4×4 | 5×5 | 6×6 | 8×8 | 10×10 | 12×12 | 15×15 | 20×20 | 25×25 | 30×30 | 35×35 |
| 1.0 | 0.610 | 0.630 | 0.633 | 0.643 | 0.667 | 0.695 | 0.722 | 0.759 | 0.810 | 0.842 | 0.853 | 0.858 |
| 2.0 | 0.885 | 0.890 | 0.889 | 0.889 | 0.900 | 0.915 | 0.929 | 0.948 | 0.969 | 0.983 | 0.986 | 0.988 |
| 3.0 | 0.975 | 0.975 | 0.974 | 0.976 | 0.978 | 0.985 | 0.989 | 0.995 | 1.004 | 1.008 | 1.007 | 1.007 |
| 4.0 | 1.000 | 1.000 | 1.000 | 1.000 | 1.000 | 1.000 | 1.000 | 1.000 | 1.000 | 1.000 | 1.000 | 1.000 |
| 5.0 | 0.965 | 0.989 | 0.994 | 0.996 | 0.995 | 0.993 | 0.991 | 0.987 | 0.986 | 0.985 | 0.984 | 0.986 |
| 6.0 | 0.940 | 0.971 | 0.980 | 0.980 | 0.979 | 0.977 | 0.974 | 0.971 | 0.967 | 0.970 | 0.967 | 0.970 |
| 8.0 | 0.902 | 0.932 | 0.935 | 0.938 | 0.939 | 0.939 | 0.937 | 0.936 | 0.934 | 0.936 | 0.936 | 0.940 |
| 10.0 | 0.855 | 0.884 | 0.889 | 0.895 | 0.896 | 0.898 | 0.899 | 0.898 | 0.898 | 0.901 | 0.904 | 0.908 |
| 12.0 | 0.794 | 0.830 | 0.842 | 0.847 | 0.854 | 0.856 | 0.858 | 0.860 | 0.862 | 0.867 | 0.870 | 0.875 |
| 15.0 | 0.723 | 0.768 | 0.780 | 0.785 | 0.794 | 0.797 | 0.803 | 0.808 | 0.813 | 0.820 | 0.824 | 0.831 |
| 18.0 | 0.650 | 0.707 | 0.720 | 0.726 | 0.736 | 0.743 | 0.749 | 0.755 | 0.762 | 0.771 | 0.777 | 0.784 |
| 20.0 | 0.615 | 0.670 | 0.682 | 0.690 | 0.700 | 0.706 | 0.712 | 0.720 | 0.730 | 0.739 | 0.744 | 0.754 |
| 23.0 | 0.562 | 0.615 | 0.629 | 0.627 | 0.647 | 0.655 | 0.663 | 0.671 | 0.683 | 0.693 | 0.701 | 0.711 |
| 25.0 | 0.530 | 0.588 | 0.598 | 0.607 | 0.617 | 0.625 | 0.633 | 0.643 | 0.654 | 0.664 | 0.673 | 0.683 |
| 28.0 | 0.487 | 0.539 | 0.552 | 0.561 | 0.571 | 0.577 | 0.587 | 0.597 | 0.610 | 0.622 | 0.631 | 0.641 |
| 30.0 | 0.455 | 0.513 | 0.523 | 0.532 | 0.544 | 0.552 | 0.559 | 0.565 | 0.583 | 0.594 | 0.603 | 0.613 |

From Johnson, D.A., Ikoro, N.C., Chang, C.-H., Scarbrough, E.C., and Antich, P.P., Properties of the 18-MV photon beam from a dual energy linear accelerator, *Med. Phys.*, 14, 1071, 1987. With permission.

**TABLE 1.E6**

**Percent Depth Doses for 18-MV X Rays [AECL Therac-20, 100 cm SSD]**

| Depth (cm) | Field Size (cm×cm) | | | | | |
|---|---|---|---|---|---|---|
| | 4×4 | 6×6 | 10×10 | 20×20 | 30×30 | 40×40 |
| 0.0 | 6.2 | 7.9 | 11.7 | 24.1 | 36.3 | 44.7 |
| 1.0 | 78.4 | 78.8 | 82.1 | 88.5 | 94.3 | 98.1 |
| 2.0 | 96.9 | 95.9 | 97.6 | 99.9 | 101.8 | 102.8 |
| 3.0 | 100.0 | 100.0 | 100.0 | 100.0 | 100.0 | 100.0 |
| 4.0 | 98.4 | 98.9 | 98.8 | 97.6 | 96.7 | 96.2 |
| 5.0 | 95.1 | 95.9 | 95.8 | 94.2 | 93.1 | 92.7 |
| 6.0 | 91.5 | 92.2 | 92.8 | 90.7 | 89.5 | 89.1 |
| 8.0 | 83.2 | 84.8 | 85.0 | 84.3 | 83.2 | 82.6 |
| 10.0 | 76.1 | 77.7 | 78.1 | 77.9 | 77.2 | 76.5 |
| 15.0 | 60.3 | 61.9 | 63.1 | 63.9 | 63.5 | 63.5 |
| 20.0 | 47.9 | 49.1 | 51.6 | 51.8 | 51.9 | 52.1 |
| 25.0 | 38.1 | 39.3 | 41.0 | 42.4 | 42.7 | 42.6 |
| 30.0 | 30.5 | 31.2 | 32.9 | 34.3 | 34.7 | 34.8 |

From Patterson, M.S. and Shragge, P.C., Characteristics of an 18 MV photon beam from a Therac 20 medical linear accelerator, *Med. Phys.*, 8, 312, 1981. With permission.

## TABLE 1.E7
### Tissue Phantom Ratios for 18-MV X Rays [AECL Therac-20, 100 cm SAD]

| Depth (cm) | Field Size (cm×cm) | | | | | | | |
|---|---|---|---|---|---|---|---|---|
| | 0×0 | 4×4 | 6×6 | 10×10 | 15×15 | 20×20 | 30×30 | 40×40 |
| 0.0 | 0.025 | 0.057 | 0.073 | 0.109 | 0.166 | 0.228 | 0.345 | 0.427 |
| 1.0 | 0.739 | 0.739 | 0.741 | 0.768 | 0.807 | 0.847 | 0.915 | 0.952 |
| 2.0 | 0.915 | 0.915 | 0.916 | 0.931 | 0.953 | 0.972 | 1.004 | 1.020 |
| 3.0 | 0.980 | 0.980 | 0.980 | 0.986 | 0.995 | 1.003 | 1.011 | 1.015 |
| 4.0 | 1.000 | 1.000 | 1.000 | 1.000 | 1.000 | 1.000 | 1.000 | 1.000 |
| 5.0 | 0.967 | 0.979 | 0.983 | 0.986 | 0.985 | 0.985 | 0.982 | 0.980 |
| 6.0 | 0.938 | 0.959 | 0.966 | 0.972 | 0.970 | 0.969 | 0.964 | 0.960 |
| 8.0 | 0.873 | 0.905 | 0.917 | 0.929 | 0.930 | 0.930 | 0.929 | 0.926 |
| 10.0 | 0.812 | 0.849 | 0.863 | 0.880 | 0.889 | 0.890 | 0.890 | 0.890 |
| 12.0 | 0.765 | 0.802 | 0.817 | 0.836 | 0.847 | 0.853 | 0.854 | 0.855 |
| 15.0 | 0.691 | 0.733 | 0.751 | 0.775 | 0.789 | 0.793 | 0.799 | 0.805 |
| 20.0 | 0.585 | 0.631 | 0.650 | 0.675 | 0.692 | 0.703 | 0.714 | 0.724 |
| 25.0 | 0.494 | 0.540 | 0.558 | 0.584 | 0.607 | 0.618 | 0.632 | 0.646 |
| 30.0 | 0.417 | 0.463 | 0.482 | 0.509 | 0.528 | 0.542 | 0.558 | 0.574 |

From Patterson, M.S. and Shragge, P.C., Characteristics of an 18 MV photon beam from a Therac 20 medical linear accelerator, *Med. Phys.*, 8, 312, 1981. With permission.

**TABLE 1.E8**
**Percent Depth Doses for 20-MV X Rays [Siemens Mevatron KD, 100 cm SSD]**

| Depth (cm) | Field Size (cm×cm) | | | | | | | | | | | | | | |
|---|---|---|---|---|---|---|---|---|---|---|---|---|---|---|---|
| | 0×0 | 2×2 | 3×3 | 4×4 | 5×5 | 8×8 | 10×10 | 12×12 | 15×15 | 20×20 | 25×25 | 30×30 | 35×35 |
| 0.0 | 3.7 | 5.3 | 6.2 | 7.0 | 7.8 | 10.6 | 12.5 | 14.3 | 17.0 | 21.5 | 25.3 | 28.0 | 30.3 |
| 1.0 | 78.2 | 75.5 | 73.5 | 71.9 | 71.4 | 72.9 | 74.6 | 76.1 | 78.3 | 81.9 | 85.1 | 88.0 | 90.7 |
| 2.0 | 95.1 | 93.9 | 93.4 | 92.8 | 92.3 | 92.7 | 93.4 | 94.3 | 95.6 | 97.5 | 98.9 | 100.1 | 101.3 |
| 3.0 | 102.3 | 102.2 | 101.7 | 100.6 | 100.1 | 100.0 | 100.2 | 100.4 | 100.9 | 101.4 | 101.7 | 101.9 | 102.1 |
| 4.0 | 100.0 | 100.0 | 100.0 | 100.0 | 100.0 | 100.0 | 100.0 | 100.0 | 100.0 | 100.0 | 100.0 | 100.0 | 100.0 |
| 5.0 | 96.1 | 96.8 | 97.4 | 97.7 | 97.8 | 97.9 | 97.8 | 97.7 | 97.5 | 97.4 | 97.2 | 97.0 | 96.8 |
| 6.0 | 92.0 | 92.7 | 93.3 | 93.7 | 93.9 | 94.2 | 94.2 | 94.2 | 94.1 | 94.0 | 93.9 | 93.7 | 93.6 |
| 7.0 | 88.0 | 88.8 | 89.4 | 89.8 | 90.1 | 90.7 | 90.8 | 90.8 | 90.8 | 90.7 | 90.6 | 90.6 | 90.5 |
| 8.0 | 81.6 | 83.5 | 84.7 | 85.7 | 86.2 | 87.1 | 87.2 | 87.3 | 87.3 | 87.4 | 87.4 | 87.4 | 87.4 |
| 10.0 | 69.5 | 73.5 | 76.1 | 77.9 | 78.9 | 80.1 | 80.4 | 80.6 | 80.8 | 81.1 | 81.3 | 81.4 | 81.6 |
| 15.0 | 57.3 | 59.8 | 61.1 | 62.1 | 63.2 | 64.8 | 65.4 | 65.8 | 66.3 | 66.9 | 67.3 | 67.5 | 67.7 |
| 20.0 | 46.7 | 48.4 | 49.3 | 50.1 | 50.8 | 52.4 | 53.1 | 53.7 | 54.3 | 55.2 | 55.9 | 56.3 | 56.8 |
| 25.0 | 36.3 | 38.5 | 39.6 | 40.6 | 41.2 | 42.7 | 43.4 | 44.0 | 44.8 | 45.7 | 46.2 | 46.5 | 46.8 |
| 30.0 | 27.4 | 30.0 | 31.4 | 42.3 | 32.8 | 34.3 | 35.1 | 35.8 | 36.6 | 37.5 | 37.7 | 37.8 | 38.0 |

From Al-Ghazi, M.S.A.L., Arjune, B., Fiedler, J.A., and Sharma, P.D., Dosimetric aspects of the therapeutic photon beams from a dual-energy linear accelerator, *Med. Phys.*, 15, 250, 1988. With permission.

## TABLE 1.E9
### Tissue Phantom Ratios for 20-MV X Rays [Siemens Mevatron KD, 100 cm SAD]

| Depth (cm) | Field Size (cm×cm) | | | | | | | | | | | | |
|---|---|---|---|---|---|---|---|---|---|---|---|---|---|
| | 0×0 | 2×2 | 3×3 | 4×4 | 5×5 | 8×8 | 10×10 | 12×12 | 15×15 | 20×20 | 25×25 | 30×30 | 35×35 |
| 0.0 | 0.034 | 0.049 | 0.057 | 0.065 | 0.072 | 0.098 | 0.116 | 0.133 | 0.157 | 0.199 | 0.234 | 0.260 | 0.281 |
| 1.0 | 0.737 | 0.712 | 0.694 | 0.679 | 0.674 | 0.687 | 0.703 | 0.717 | 0.738 | 0.772 | 0.802 | 0.829 | 0.854 |
| 2.0 | 0.915 | 0.903 | 0.899 | 0.893 | 0.888 | 0.891 | 0.898 | 0.906 | 0.919 | 0.937 | 0.951 | 0.962 | 0.973 |
| 3.0 | 1.022 | 1.013 | 0.990 | 0.988 | 0.983 | 0.981 | 0.983 | 0.985 | 0.989 | 0.995 | 0.998 | 1.000 | 1.001 |
| 4.0 | 1.000 | 1.000 | 1.000 | 1.000 | 1.000 | 1.000 | 1.000 | 1.000 | 1.000 | 1.000 | 1.000 | 1.000 | 1.000 |
| 5.0 | 0.980 | 0.986 | 0.992 | 0.996 | 0.997 | 0.997 | 0.997 | 0.996 | 0.994 | 0.993 | 0.991 | 0.989 | 0.987 |
| 6.0 | 0.956 | 0.963 | 0.968 | 0.972 | 0.975 | 0.979 | 0.979 | 0.978 | 0.977 | 0.976 | 0.975 | 0.974 | 0.972 |
| 7.0 | 0.931 | 0.939 | 0.945 | 0.949 | 0.952 | 0.960 | 0.961 | 0.961 | 0.961 | 0.960 | 0.959 | 0.958 | 0.957 |
| 8.0 | 0.880 | 0.898 | 0.911 | 0.921 | 0.928 | 0.938 | 0.940 | 0.941 | 0.941 | 0.942 | 0.942 | 0.942 | 0.942 |
| 10.0 | 0.778 | 0.817 | 0.844 | 0.865 | 0.878 | 0.894 | 0.898 | 0.900 | 0.902 | 0.905 | 0.908 | 0.909 | 0.910 |
| 15.0 | 0.701 | 0.727 | 0.741 | 0.755 | 0.766 | 0.787 | 0.795 | 0.800 | 0.805 | 0.814 | 0.819 | 0.822 | 0.824 |
| 20.0 | 0.622 | 0.640 | 0.650 | 0.660 | 0.669 | 0.689 | 0.699 | 0.706 | 0.714 | 0.726 | 0.736 | 0.741 | 0.742 |
| 25.0 | 0.525 | 0.549 | 0.563 | 0.576 | 0.586 | 0.606 | 0.616 | 0.624 | 0.634 | 0.649 | 0.659 | 0.664 | 0.666 |
| 30.0 | 0.428 | 0.458 | 0.476 | 0.492 | 0.503 | 0.523 | 0.534 | 0.543 | 0.555 | 0.572 | 0.582 | 0.586 | 0.587 |

From Al-Ghazi, M.S.A.L., Arjune, B., Fiedler, J.A., and Sharma, P.D., Dosimetric aspects of the therapeutic photon beams from a dual-energy linear accelerator, *Med. Phys.*, 15, 250, 1988. With permission.

**TABLE 1.F1**
**Percent Depth Doses for 23-MV X Rays [CGR Saturne-25, 100 cm SSD]**

| Depth (cm) | Field Size (cm×cm) | | | | | | | | | | | | | |
|---|---|---|---|---|---|---|---|---|---|---|---|---|---|---|
| | 4×4 | 5×5 | 6×6 | 8×8 | 10×10 | 12×12 | 14×14 | 16×16 | 18×18 | 20×20 | 25×25 | 30×30 | 35×35 | 40×40 |
| 0.0 | 5.5 | 6.3 | 7.3 | 9.7 | 12.2 | 14.9 | 17.7 | 20.5 | 23.4 | 26.3 | 33.1 | 38.9 | 44.0 | 46.5 |
| 0.1 | 18.4 | 19.0 | 20.1 | 22.4 | 24.9 | 27.8 | 30.7 | 33.7 | 36.6 | 39.5 | 46.5 | 52.2 | 56.8 | 59.1 |
| 0.5 | 43.5 | 43.7 | 43.9 | 46.3 | 49.9 | 51.2 | 54.3 | 57.2 | 59.9 | 62.7 | 68.5 | 73.8 | 77.5 | 79.4 |
| 1.0 | 67.0 | 67.4 | 67.8 | 69.0 | 70.5 | 73.8 | 74.8 | 76.6 | 79.3 | 82.0 | 85.5 | 89.0 | 90.4 | 91.0 |
| 2.0 | 91.3 | 90.4 | 90.8 | 91.4 | 92.0 | 93.6 | 94.1 | 94.8 | 95.5 | 96.3 | 98.3 | 99.3 | 99.6 | 99.7 |
| 3.0 | 98.5 | 98.3 | 98.3 | 98.5 | 99.0 | 99.4 | 99.5 | 99.6 | 99.8 | 100.0 | 99.9 | 99.7 | 99.3 | 99.3 |
| 4.0 | 99.9 | 100.0 | 100.0 | 99.9 | 99.9 | 99.8 | 99.7 | 99.5 | 99.2 | 98.9 | 98.2 | 97.3 | 96.8 | 96.8 |
| 5.0 | 98.0 | 98.5 | 98.8 | 98.5 | 98.2 | 98.0 | 97.6 | 97.2 | 96.7 | 96.2 | 95.2 | 94.0 | 93.5 | 93.7 |
| 6.0 | 95.0 | 95.7 | 96.0 | 95.7 | 95.4 | 95.1 | 94.7 | 94.2 | 93.7 | 93.2 | 92.0 | 90.9 | 90.4 | 90.7 |
| 8.0 | 87.7 | 88.5 | 89.0 | 88.9 | 88.6 | 88.4 | 88.1 | 87.7 | 87.2 | 86.6 | 85.7 | 84.6 | 84.0 | 84.5 |
| 10.0 | 80.5 | 81.3 | 82.0 | 82.1 | 82.0 | 81.7 | 81.4 | 81.1 | 80.7 | 80.3 | 79.5 | 78.6 | 78.3 | 78.8 |
| 12.0 | 74.0 | 74.7 | 75.4 | 75.7 | 75.8 | 75.7 | 75.6 | 75.3 | 75.0 | 74.6 | 73.9 | 73.0 | 73.0 | 73.5 |
| 14.0 | 67.8 | 68.6 | 69.3 | 69.7 | 69.9 | 69.8 | 69.8 | 69.7 | 69.4 | 69.2 | 68.5 | 67.9 | 67.9 | 68.3 |
| 16.0 | 62.2 | 62.8 | 63.7 | 64.1 | 64.5 | 64.6 | 64.6 | 64.5 | 64.3 | 64.1 | 63.6 | 63.0 | 63.2 | 63.6 |
| 18.0 | 57.0 | 57.8 | 58.5 | 59.1 | 59.3 | 59.7 | 59.8 | 59.8 | 59.6 | 59.4 | 59.0 | 58.6 | 58.7 | 59.2 |
| 20.0 | 52.3 | 53.0 | 53.8 | 54.4 | 54.8 | 55.2 | 55.3 | 55.2 | 55.1 | 55.0 | 54.9 | 54.4 | 54.6 | 54.9 |
| 22.0 | 48.1 | 48.8 | 49.5 | 50.1 | 50.7 | 51.0 | 51.1 | 51.1 | 51.1 | 51.1 | 50.8 | 50.6 | 50.8 | 51.1 |
| 24.0 | 44.3 | 44.9 | 45.7 | 46.2 | 46.8 | 47.0 | 47.3 | 47.4 | 47.3 | 47.3 | 47.2 | 46.9 | 47.2 | 47.4 |
| 25.0 | 42.5 | 43.1 | 43.8 | 44.4 | 44.9 | 45.3 | 45.5 | 45.6 | 45.5 | 45.5 | 45.5 | 45.4 | 45.3 | 45.5 |
| 26.0 | 40.7 | 41.3 | 42.0 | 42.6 | 43.2 | 43.6 | 43.8 | 43.9 | 43.9 | 43.9 | 43.8 | 43.6 | 43.8 | 43.9 |
| 28.0 | 37.5 | 38.0 | 38.6 | 39.3 | 39.8 | 40.2 | 40.5 | 40.6 | 40.6 | 40.6 | 40.7 | 40.5 | 40.7 | 40.8 |
| 30.0 | 34.5 | 34.9 | 35.5 | 36.2 | 36.5 | 36.9 | 37.2 | 37.4 | 37.4 | 37.5 | 37.8 | 37.7 | 37.9 | 38.1 |

From Luxton, G. and Astrahan, M.A., Characteristics of the high-energy photon beam of a 25-MeV accelerator, *Med. Phys.*, 15, 82, 1988. With permission.

## TABLE 1.F2
## Tissue Phantom Ratios for 23-MV X Rays [CGR Saturne-25, 100 cm SAD]

| Depth (cm) | Field Size (cm×cm) | | | | | | | | | | | | |
|---|---|---|---|---|---|---|---|---|---|---|---|---|---|
| | 4×4 | 5×5 | 6×6 | 8×8 | 10×10 | 12×12 | 14×14 | 16×16 | 18×18 | 20×20 | 25×25 | 30×30 | 40×40 |
| 0.0 | 0.051 | 0.058 | 0.067 | 0.089 | 0.113 | 0.138 | 0.165 | 0.192 | 0.220 | 0.248 | 0.315 | 0.375 | 0.451 |
| 0.1 | 0.171 | 0.176 | 0.185 | 0.207 | 0.231 | 0.258 | 0.286 | 0.315 | 0.344 | 0.373 | 0.444 | 0.504 | 0.574 |
| 0.5 | 0.407 | 0.407 | 0.408 | 0.430 | 0.465 | 0.479 | 0.509 | 0.538 | 0.567 | 0.596 | 0.658 | 0.717 | 0.777 |
| 1.0 | 0.634 | 0.634 | 0.636 | 0.648 | 0.664 | 0.695 | 0.709 | 0.728 | 0.757 | 0.786 | 0.829 | 0.873 | 0.900 |
| 2.0 | 0.881 | 0.868 | 0.868 | 0.875 | 0.884 | 0.900 | 0.909 | 0.919 | 0.931 | 0.943 | 0.971 | 0.994 | 1.005 |
| 3.0 | 0.969 | 0.962 | 0.959 | 0.962 | 0.969 | 0.975 | 0.980 | 0.985 | 0.992 | 0.998 | 1.008 | 1.018 | 1.021 |
| 4.0 | 1.001 | 0.997 | 0.994 | 0.995 | 0.998 | 0.999 | 1.001 | 1.004 | 1.006 | 1.008 | 1.011 | 1.015 | 1.015 |
| 5.0 | 1.000 | 1.000 | 1.000 | 1.000 | 1.000 | 1.000 | 1.000 | 1.000 | 1.000 | 1.000 | 1.000 | 1.000 | 1.000 |
| 6.0 | 0.987 | 0.989 | 0.990 | 0.990 | 0.990 | 0.989 | 0.989 | 0.989 | 0.988 | 0.988 | 0.986 | 0.985 | 0.986 |
| 8.0 | 0.944 | 0.948 | 0.951 | 0.954 | 0.954 | 0.954 | 0.954 | 0.956 | 0.955 | 0.953 | 0.952 | 0.953 | 0.951 |
| 10.0 | 0.897 | 0.903 | 0.906 | 0.913 | 0.915 | 0.915 | 0.915 | 0.916 | 0.916 | 0.916 | 0.916 | 0.917 | 0.919 |
| 12.0 | 0.853 | 0.859 | 0.862 | 0.872 | 0.875 | 0.878 | 0.879 | 0.881 | 0.882 | 0.882 | 0.883 | 0.884 | 0.887 |
| 14.0 | 0.808 | 0.816 | 0.820 | 0.831 | 0.835 | 0.838 | 0.839 | 0.843 | 0.845 | 0.846 | 0.848 | 0.850 | 0.854 |
| 16.0 | 0.766 | 0.773 | 0.776 | 0.790 | 0.795 | 0.801 | 0.804 | 0.807 | 0.810 | 0.810 | 0.814 | 0.817 | 0.822 |
| 18.0 | 0.726 | 0.734 | 0.738 | 0.752 | 0.757 | 0.762 | 0.768 | 0.773 | 0.777 | 0.777 | 0.780 | 0.784 | 0.790 |
| 20.0 | 0.688 | 0.695 | 0.699 | 0.715 | 0.721 | 0.727 | 0.733 | 0.738 | 0.741 | 0.743 | 0.748 | 0.754 | 0.759 |
| 22.0 | 0.653 | 0.660 | 0.664 | 0.679 | 0.686 | 0.694 | 0.700 | 0.704 | 0.708 | 0.710 | 0.717 | 0.721 | 0.729 |
| 24.0 | 0.620 | 0.628 | 0.631 | 0.647 | 0.652 | 0.661 | 0.666 | 0.671 | 0.677 | 0.680 | 0.686 | 0.691 | 0.698 |
| 25.0 | 0.603 | 0.612 | 0.615 | 0.630 | 0.637 | 0.644 | 0.651 | 0.656 | 0.662 | 0.664 | 0.670 | 0.677 | 0.684 |
| 26.0 | 0.588 | 0.595 | 0.598 | 0.614 | 0.620 | 0.629 | 0.636 | 0.642 | 0.647 | 0.650 | 0.657 | 0.662 | 0.669 |
| 28.0 | 0.557 | 0.565 | 0.568 | 0.582 | 0.590 | 0.597 | 0.604 | 0.610 | 0.616 | 0.620 | 0.626 | 0.634 | 0.641 |
| 30.0 | 0.527 | 0.535 | 0.539 | 0.552 | 0.560 | 0.567 | 0.571 | 0.577 | 0.583 | 0.589 | 0.595 | 0.603 | 0.615 |

From Luxton, G. and Astrahan, M.A., Characteristics of the high-energy photon beam of a 25-MeV accelerator, *Med. Phys.*, 15, 82, 1988. With permission.

TABLE 1.F3
Percent Depth Doses for 24-MV X Rays [Varian Clinac-2500, 100 cm SSD]

| Depth (cm) | Field Size (cm×cm) | | | | | | | | | | |
|---|---|---|---|---|---|---|---|---|---|---|---|
| | 4×4 | 5×5 | 6×6 | 8×8 | 10×10 | 12×12 | 15×15 | 20×20 | 25×25 | 30×30 | 36×36 |
| 1.0 | 65.6 | 65.5 | 66.8 | 70.5 | 74.6 | 77.8 | 82.6 | 88.1 | 90.8 | 92.5 | 93.1 |
| 2.0 | 89.3 | 89.0 | 89.6 | 91.5 | 93.9 | 96.1 | 98.8 | 101.3 | 102.4 | 103.2 | 103.5 |
| 3.0 | 98.3 | 98.3 | 98.7 | 99.6 | 100.5 | 101.6 | 102.9 | 104.2 | 104.6 | 104.8 | 104.9 |
| 4.0 | 100.9 | 100.8 | 100.8 | 101.0 | 101.4 | 101.8 | 102.3 | 102.9 | 103.0 | 103.0 | 103.0 |
| 5.0 | 100.0 | 100.0 | 100.0 | 100.0 | 100.0 | 100.0 | 100.0 | 100.0 | 100.0 | 100.0 | 100.0 |
| 6.0 | 97.0 | 97.1 | 97.1 | 97.0 | 96.9 | 96.9 | 96.8 | 96.8 | 96.8 | 96.9 | 86.9 |
| 7.0 | 94.0 | 94.3 | 94.2 | 94.0 | 93.9 | 93.8 | 93.6 | 93.7 | 93.7 | 93.8 | 93.9 |
| 8.0 | 90.4 | 90.6 | 90.6 | 90.5 | 90.5 | 90.5 | 90.3 | 90.5 | 90.6 | 90.8 | 90.9 |
| 9.0 | 86.8 | 87.1 | 87.1 | 87.2 | 87.2 | 87.2 | 87.1 | 87.4 | 87.6 | 87.8 | 88.0 |
| 10.0 | 83.3 | 83.6 | 83.8 | 83.9 | 84.0 | 84.0 | 84.1 | 84.4 | 84.7 | 85.0 | 85.2 |
| 12.0 | 76.7 | 77.0 | 77.2 | 77.5 | 77.7 | 77.8 | 78.0 | 78.5 | 79.0 | 79.3 | 79.7 |
| 14.0 | 70.5 | 70.7 | 71.1 | 71.5 | 71.8 | 72.0 | 72.3 | 73.1 | 73.6 | 74.1 | 74.5 |
| 16.0 | 64.8 | 65.0 | 65.5 | 65.9 | 66.3 | 66.6 | 67.0 | 68.0 | 68.6 | 69.1 | 69.6 |
| 18.0 | 59.6 | 59.9 | 60.4 | 60.8 | 61.3 | 61.6 | 62.1 | 63.1 | 63.8 | 64.4 | 64.8 |
| 20.0 | 54.8 | 55.0 | 55.6 | 56.0 | 56.5 | 56.9 | 57.5 | 58.5 | 59.3 | 59.9 | 60.4 |
| 22.0 | 50.5 | 50.8 | 51.3 | 51.8 | 52.3 | 52.7 | 53.3 | 54.4 | 55.2 | 55.8 | 56.3 |
| 24.0 | 46.6 | 46.8 | 47.3 | 47.8 | 48.3 | 48.6 | 49.3 | 50.5 | 51.4 | 51.9 | 52.4 |
| 26.0 | 43.1 | 43.4 | 43.8 | 44.2 | 44.7 | 45.1 | 45.8 | 46.9 | 47.8 | 48.4 | 48.8 |
| 28.0 | 39.8 | 40.1 | 40.5 | 40.9 | 41.4 | 41.8 | 42.5 | 43.5 | 44.4 | 45.0 | 45.4 |
| 30.0 | 36.7 | 37.0 | 37.4 | 37.8 | 38.2 | 38.7 | 39.3 | 40.3 | 41.2 | 41.8 | 42.3 |

Adapted from The University of Iowa Hospitals and Clinics, Iowa City, IA, 1992, personal communication.

## TABLE 1.F4
### Tissue Phantom Ratios for 24-MV X Rays [Varian Clinac-2500, 100 cm SAD]

| Depth (cm) | Field Size (cm×cm) | | | | | | | | | | |
|---|---|---|---|---|---|---|---|---|---|---|---|
| | 4×4 | 5×5 | 6×6 | 8×8 | 10×10 | 12×12 | 15×15 | 20×20 | 25×25 | 30×30 | 36×36 |
| 1.0 | 0.610 | 0.608 | 0.621 | 0.651 | 0.689 | 0.718 | 0.762 | 0.814 | 0.839 | 0.855 | 0.862 |
| 2.0 | 0.846 | 0.842 | 0.848 | 0.862 | 0.884 | 0.904 | 0.930 | 0.955 | 0.966 | 0.973 | 0.976 |
| 3.0 | 0.949 | 0.947 | 0.950 | 0.957 | 0.966 | 0.976 | 0.988 | 1.002 | 1.006 | 1.008 | 1.009 |
| 4.0 | 0.991 | 0.990 | 0.990 | 0.991 | 0.995 | 0.998 | 1.002 | 1.009 | 1.010 | 1.010 | 1.010 |
| 5.0 | 1.000 | 1.000 | 1.000 | 1.000 | 1.000 | 1.000 | 1.000 | 1.000 | 1.000 | 1.000 | 1.000 |
| 6.0 | 0.987 | 0.988 | 0.988 | 0.988 | 0.988 | 0.987 | 0.986 | 0.986 | 0.987 | 0.987 | 0.987 |
| 7.0 | 0.974 | 0.977 | 0.977 | 0.977 | 0.976 | 0.975 | 0.973 | 0.972 | 0.973 | 0.974 | 0.975 |
| 8.0 | 0.952 | 0.955 | 0.956 | 0.958 | 0.958 | 0.957 | 0.956 | 0.956 | 0.958 | 0.960 | 0.961 |
| 9.0 | 0.930 | 0.934 | 0.936 | 0.939 | 0.939 | 0.940 | 0.939 | 0.940 | 0.943 | 0.945 | 0.948 |
| 10.0 | 0.907 | 0.912 | 0.915 | 0.920 | 0.921 | 0.922 | 0.923 | 0.924 | 0.928 | 0.931 | 0.934 |
| 12.0 | 0.864 | 0.869 | 0.873 | 0.879 | 0.882 | 0.884 | 0.886 | 0.891 | 0.896 | 0.900 | 0.904 |
| 14.0 | 0.820 | 0.825 | 0.830 | 0.839 | 0.844 | 0.847 | 0.850 | 0.857 | 0.864 | 0.869 | 0.875 |
| 16.0 | 0.779 | 0.783 | 0.790 | 0.800 | 0.806 | 0.810 | 0.814 | 0.824 | 0.832 | 0.838 | 0.845 |
| 18.0 | 0.740 | 0.744 | 0.751 | 0.763 | 0.769 | 0.774 | 0.779 | 0.790 | 0.799 | 0.806 | 0.813 |
| 20.0 | 0.701 | 0.706 | 0.713 | 0.725 | 0.732 | 0.738 | 0.745 | 0.756 | 0.766 | 0.774 | 0.782 |
| 22.0 | 0.668 | 0.672 | 0.679 | 0.692 | 0.699 | 0.705 | 0.712 | 0.724 | 0.736 | 0.745 | 0.753 |
| 24.0 | 0.634 | 0.637 | 0.645 | 0.658 | 0.665 | 0.672 | 0.679 | 0.692 | 0.705 | 0.715 | 0.723 |
| 26.0 | 0.603 | 0.608 | 0.616 | 0.628 | 0.636 | 0.642 | 0.649 | 0.663 | 0.675 | 0.686 | 0.694 |
| 28.0 | 0.573 | 0.578 | 0.586 | 0.599 | 0.606 | 0.612 | 0.620 | 0.633 | 0.646 | 0.656 | 0.666 |
| 30.0 | 0.542 | 0.548 | 0.556 | 0.569 | 0.576 | 0.583 | 0.591 | 0.604 | 0.616 | 0.627 | 0.637 |

Adapted from The University of Iowa Hospitals and Clinics, Iowa City, IA, 1992, personal communication.

## TABLE 1.F5
### Percent Depth Doses for 25-MV X Rays [Varian CL-35, 100 cm SSD]

| Depth (cm) | Field Size (cm×cm) | | | | | | | | | | | | | | |
|---|---|---|---|---|---|---|---|---|---|---|---|---|---|---|---|
| | 4×4 | 6×6 | 8×8 | 10×10 | 12×12 | 14×14 | 16×16 | 18×18 | 20×20 | 22×22 | 24×24 | 26×26 | 28×28 | 30×30 | 35×35 |
| 1.0 | 74.2 | 76.6 | 78.0 | 78.8 | 84.5 | 85.8 | 88.3 | 89.4 | 90.8 | 90.2 | 92.0 | 92.5 | 93.1 | 93.4 | 93.6 |
| 2.0 | 93.5 | 93.3 | 94.6 | 94.7 | 97.5 | 98.0 | 99.0 | 99.2 | 99.4 | 99.5 | 99.8 | 99.9 | 99.9 | 99.9 | 99.9 |
| 3.0 | 99.2 | 99.4 | 99.6 | 99.8 | 100.0 | 100.0 | 99.6 | 99.7 | 99.8 | 99.7 | 99.6 | 99.5 | 99.5 | 99.5 | 99.3 |
| 4.0 | 100.0 | 100.0 | 99.8 | 99.5 | 99.1 | 98.5 | 98.1 | 97.6 | 97.5 | 97.4 | 97.2 | 96.8 | 96.5 | 96.7 | 97.3 |
| 5.0 | 99.0 | 98.5 | 98.0 | 97.5 | 96.5 | 96.0 | 95.4 | 94.5 | 94.0 | 93.5 | 93.1 | 93.0 | 93.0 | 93.0 | 93.2 |
| 6.0 | 96.2 | 95.5 | 94.8 | 94.2 | 93.5 | 92.7 | 91.9 | 91.3 | 90.7 | 90.3 | 90.0 | 89.7 | 89.7 | 89.8 | 89.8 |
| 7.0 | 93.0 | 92.5 | 91.9 | 91.1 | 90.4 | 89.5 | 88.7 | 87.8 | 87.0 | 86.4 | 86.2 | 86.2 | 86.2 | 86.3 | 86.4 |
| 8.0 | 90.0 | 89.3 | 88.3 | 87.5 | 86.8 | 85.8 | 85.0 | 84.5 | 83.8 | 83.3 | 83.0 | 83.1 | 83.1 | 83.2 | 83.5 |
| 9.0 | 86.2 | 85.7 | 84.7 | 84.0 | 83.5 | 82.4 | 81.6 | 81.1 | 80.8 | 80.3 | 80.0 | 80.1 | 80.1 | 80.1 | 80.4 |
| 10.0 | 83.0 | 82.4 | 81.8 | 81.2 | 80.7 | 79.7 | 79.0 | 78.4 | 78.0 | 77.5 | 77.4 | 77.4 | 77.5 | 77.5 | 77.7 |
| 11.0 | 79.5 | 79.1 | 78.4 | 77.9 | 77.4 | 76.7 | 76.0 | 75.5 | 75.2 | 75.1 | 75.0 | 75.0 | 75.1 | 75.1 | 75.1 |
| 12.0 | 76.1 | 75.8 | 75.4 | 75.0 | 74.8 | 73.5 | 73.5 | 73.0 | 72.5 | 72.5 | 72.4 | 72.4 | 72.4 | 72.5 | 72.5 |
| 13.0 | 73.1 | 72.9 | 72.7 | 72.5 | 71.9 | 71.3 | 70.9 | 70.4 | 70.0 | 70.0 | 69.9 | 69.9 | 69.9 | 70.0 | 70.0 |
| 14.0 | 70.1 | 69.9 | 69.7 | 69.4 | 69.0 | 68.7 | 68.3 | 67.8 | 67.5 | 67.5 | 67.4 | 67.4 | 67.4 | 67.5 | 67.5 |
| 15.0 | 67.2 | 67.2 | 67.0 | 66.9 | 66.7 | 66.5 | 65.9 | 65.5 | 65.2 | 64.8 | 64.9 | 64.9 | 64.9 | 65.0 | 65.0 |
| 16.0 | 64.5 | 64.5 | 64.3 | 64.3 | 64.3 | 63.9 | 63.2 | 63.2 | 62.8 | 62.5 | 62.5 | 62.5 | 62.5 | 62.5 | 62.9 |
| 17.0 | 61.8 | 61.8 | 61.8 | 61.7 | 61.7 | 61.5 | 61.0 | 60.9 | 60.5 | 60.2 | 60.3 | 60.3 | 60.3 | 60.4 | 60.7 |
| 18.0 | 59.3 | 59.3 | 59.3 | 59.3 | 59.3 | 59.2 | 58.7 | 58.5 | 58.4 | 58.2 | 58.2 | 58.2 | 58.2 | 58.3 | 58.6 |
| 19.0 | 56.8 | 56.8 | 56.8 | 56.8 | 56.8 | 56.8 | 56.5 | 56.2 | 56.2 | 56.1 | 56.0 | 56.0 | 56.1 | 56.1 | 56.7 |
| 20.0 | 54.6 | 54.6 | 54.6 | 54.6 | 54.6 | 54.6 | 54.4 | 54.2 | 54.2 | 54.0 | 54.0 | 54.0 | 54.0 | 54.1 | 54.6 |
| 21.0 | 52.5 | 52.5 | 52.5 | 52.5 | 52.5 | 52.5 | 52.3 | 52.0 | 52.0 | 52.0 | 52.0 | 52.0 | 52.0 | 52.0 | 52.7 |
| 22.0 | 50.5 | 50.5 | 50.5 | 50.5 | 50.5 | 50.5 | 50.4 | 50.3 | 50.3 | 50.2 | 50.1 | 50.1 | 50.1 | 50.1 | 51.0 |
| 23.0 | 48.6 | 48.6 | 48.6 | 48.6 | 48.6 | 48.6 | 48.5 | 48.4 | 48.4 | 48.4 | 48.4 | 48.4 | 48.4 | 48.5 | 49.2 |
| 24.0 | 46.8 | 46.8 | 46.6 | 46.5 | 46.5 | 46.5 | 46.5 | 46.5 | 46.5 | 46.5 | 46.5 | 46.5 | 46.5 | 46.8 | 47.3 |
| 25.0 | 45.0 | 45.0 | 45.0 | 45.0 | 45.0 | 45.0 | 45.0 | 45.0 | 45.0 | 45.0 | 45.0 | 45.0 | 45.1 | 45.2 | 45.8 |

From Purdy, J.A., Keys, D.J., and Abrath, F.G., 25 MV x-ray beam characteristics from a 35 MeV linear accelerator, *Int. J. Radiat. Oncol. Biol. Phys.*, 4, 337, 1978. With permission.

## TABLE 1.F6
### Tissue Phantom Ratios for 25-MV X Rays [Varian Clinac-35, 100 cm SAD]

| Depth (cm) | Field Size (cm×cm) | | | | | | | | | | | |
|---|---|---|---|---|---|---|---|---|---|---|---|---|
| | 4×4 | 6×6 | 8×8 | 10×10 | 12×12 | 14×14 | 16×16 | 18×18 | 20×20 | 25×25 | 30×30 | 35×35 |
| 1.0 | 0.612 | 0.642 | 0.669 | 0.691 | 0.711 | 0.730 | 0.748 | 0.763 | 0.776 | 0.815 | 0.846 | 0.881 |
| 2.0 | 0.862 | 0.879 | 0.898 | 0.916 | 0.934 | 0.948 | 0.960 | 0.972 | 0.981 | 0.998 | 1.007 | 1.010 |
| 3.0 | 0.958 | 0.969 | 0.978 | 0.988 | 0.997 | 1.005 | 1.010 | 1.012 | 1.017 | 1.023 | 1.026 | 1.030. |
| 4.0 | 0.995 | 0.997 | 1.001 | 1.005 | 1.007 | 1.010 | 1.013 | 1.016 | 1.020 | 1.023 | 1.023 | 1.020 |
| 5.0 | 1.000 | 1.000 | 1.000 | 1.000 | 1.000 | 1.000 | 1.000 | 1.000 | 1.000 | 1.000 | 1.000 | 1.000 |
| 6.0 | 0.985 | 0.988 | 0.988 | 0.987 | 0.987 | 0.986 | 0.985 | 0.985 | 0.984 | 0.982 | 0.980 | 0.978 |
| 7.0 | 0.964 | 0.977 | 0.975 | 0.971 | 0.967 | 0.963 | 0.960 | 0.957 | 0.956 | 0.956 | 0.956 | 0.957 |
| 8.0 | 0.941 | 0.953 | 0.951 | 0.950 | 0.948 | 0.945 | 0.941 | 0.938 | 0.938 | 0.937 | 0.938 | 0.940 |
| 9.0 | 0.922 | 0.935 | 0.935 | 0.933 | 0.930 | 0.927 | 0.925 | 0.922 | 0.921 | 0.921 | 0.922 | 0.924 |
| 10.0 | 0.895 | 0.911 | 0.910 | 0.909 | 0.907 | 0.905 | 0.903 | 0.901 | 0.900 | 0.902 | 0.905 | 0.908 |
| 11.0 | 0.870 | 0.888 | 0.888 | 0.888 | 0.888 | 0.887 | 0.884 | 0.881 | 0.880 | 0.883 | 0.887 | 0.891 |
| 12.0 | 0.847 | 0.866 | 0.866 | 0.868 | 0.868 | 0.868 | 0.866 | 0.865 | 0.863 | 0.868 | 0.871 | 0.875 |
| 13.0 | 0.828 | 0.840 | 0.844 | 0.846 | 0.850 | 0.851 | 0.850 | 0.848 | 0.845 | 0.850 | 0.854 | 0.858 |
| 14.0 | 0.805 | 0.821 | 0.824 | 0.826 | 0.828 | 0.830 | 0.829 | 0.828 | 0.827 | 0.832 | 0.836 | 0.840 |
| 15.0 | 0.783 | 0.802 | 0.805 | 0.807 | 0.810 | 0.810 | 0.811 | 0.812 | 0.812 | 0.816 | 0.821 | 0.826 |
| 16.0 | 0.760 | 0.781 | 0.784 | 0.787 | 0.789 | 0.791 | 0.793 | 0.794 | 0.796 | 0.802 | 0.807 | 0.811 |
| 17.0 | 0.740 | 0.763 | 0.767 | 0.770 | 0.773 | 0.775 | 0.776 | 0.778 | 0.779 | 0.784 | 0.790 | 0.795 |
| 18.0 | 0.720 | 0.743 | 0.748 | 0.752 | 0.755 | 0.757 | 0.759 | 0.762 | 0.765 | 0.769 | 0.775 | 0.780 |
| 19.0 | 0.700 | 0.724 | 0.730 | 0.734 | 0.737 | 0.741 | 0.744 | 0.747 | 0.749 | 0.754 | 0.760 | 0.765 |
| 20.0 | 0.682 | 0.704 | 0.710 | 0.715 | 0.718 | 0.722 | 0.726 | 0.729 | 0.731 | 0.738 | 0.744 | 0.747 |

From Purdy, J.A., Keys, D.J., and Abrath, F.G., 25 MV x-ray beam characteristics from a 35 MeV linear accelerator, *Int. J. Radiat. Oncol. Biol. Phys.*, 4, 337, 1978. With permission.

## TABLE 1.F7

## Percent Depth Doses for 25-MV X Rays [Philips SL-25, 100 cm SSD]

| Depth (cm) | Field Size (cm×cm) | | | | | | | | | | |
|---|---|---|---|---|---|---|---|---|---|---|---|
| | 4×4 | 6×6 | 8×8 | 10×10 | 12×12 | 15×15 | 20×20 | 25×25 | 30×30 | 35×35 | 40×40 |
| 1.0 | 63.1 | 63.8 | 67.1 | 70.9 | 74.5 | 78.6 | 83.7 | 86.8 | 88.5 | 89.3 | 89.5 |
| 2.0 | 89.1 | 89.3 | 91.1 | 92.8 | 94.3 | 96.2 | 98.3 | 99.4 | 99.8 | 99.9 | 99.8 |
| 3.0 | 98.6 | 98.7 | 99.0 | 99.5 | 99.7 | 100.2 | 100.6 | 100.8 | 100.0 | 100.0 | 100.9 |
| 3.5 | 100.0 | 100.0 | 100.0 | 100.0 | 100.0 | 100.0 | 100.0 | 100.0 | 100.0 | 100.0 | 100.0 |
| 4.0 | 99.8 | 99.9 | 99.9 | 99.7 | 99.5 | 99.1 | 98.3 | 97.9 | 97.9 | 97.8 | 97.8 |
| 5.0 | 98.0 | 98.3 | 98.0 | 97.6 | 97.1 | 96.3 | 95.3 | 94.8 | 94.8 | 94.8 | 94.8 |
| 6.0 | 95.2 | 95.4 | 95.0 | 94.5 | 93.8 | 93.1 | 92.1 | 91.8 | 91.8 | 91.9 | 92.0 |
| 7.0 | 91.4 | 91.9 | 91.6 | 91.2 | 90.7 | 89.9 | 89.1 | 88.8 | 88.8 | 88.9 | 89.0 |
| 8.0 | 87.8 | 88.5 | 88.3 | 87.9 | 87.6 | 86.8 | 86.1 | 85.8 | 85.9 | 86.0 | 86.1 |
| 9.0 | 84.5 | 85.1 | 85.0 | 84.7 | 84.3 | 83.7 | 83.1 | 82.9 | 82.9 | 83.1 | 83.2 |
| 10.0 | 81.3 | 81.8 | 81.7 | 81.5 | 81.2 | 80.7 | 80.2 | 80.0 | 80.2 | 80.3 | 80.4 |
| 11.0 | 78.0 | 78.7 | 78.6 | 78.4 | 78.2 | 77.8 | 77.3 | 77.2 | 77.3 | 77.6 | 77.7 |
| 12.0 | 74.8 | 75.6 | 75.7 | 75.5 | 75.3 | 75.0 | 74.6 | 74.5 | 74.7 | 74.9 | 75.0 |
| 13.0 | 71.8 | 72.5 | 72.7 | 72.6 | 72.5 | 72.3 | 71.9 | 71.9 | 72.1 | 72.3 | 72.4 |
| 14.0 | 68.9 | 69.6 | 69.8 | 69.8 | 69.8 | 69.6 | 69.3 | 69.3 | 69.6 | 69.9 | 70.0 |
| 15.0 | 66.1 | 66.8 | 67.0 | 67.1 | 67.1 | 66.9 | 66.8 | 67.0 | 67.2 | 67.5 | 67.6 |
| 16.0 | 63.4 | 64.1 | 64.3 | 64.5 | 64.6 | 64.5 | 64.4 | 64.6 | 64.8 | 65.1 | 65.1 |
| 17.0 | 60.8 | 61.5 | 61.8 | 62.0 | 62.1 | 62.1 | 62.1 | 62.2 | 62.5 | 62.7 | 62.8 |
| 18.0 | 58.3 | 59.0 | 59.3 | 59.6 | 59.7 | 59.8 | 59.8 | 60.0 | 60.3 | 60.5 | 60.5 |
| 19.0 | 56.0 | 56.7 | 57.0 | 57.3 | 57.4 | 57.5 | 57.6 | 57.9 | 58.2 | 58.4 | 58.5 |
| 20.0 | 53.6 | 54.4 | 54.8 | 55.1 | 55.2 | 55.3 | 55.5 | 55.8 | 56.2 | 56.3 | 56.3 |
| 21.0 | 51.5 | 52.3 | 52.8 | 53.1 | 53.2 | 53.3 | 53.5 | 53.8 | 54.2 | 54.4 | 54.4 |
| 22.0 | 49.4 | 50.2 | 50.8 | 51.0 | 51.2 | 51.4 | 51.6 | 51.9 | 52.3 | 52.5 | 52.5 |
| 23.0 | 47.5 | 48.2 | 48.8 | 49.1 | 49.3 | 49.5 | 49.7 | 50.1 | 50.5 | 50.7 | 50.6 |
| 24.0 | 45.5 | 46.3 | 46.9 | 47.3 | 47.5 | 47.7 | 48.0 | 48.3 | 48.7 | 48.9 | 48.9 |
| 25.0 | 43.8 | 44.4 | 45.0 | 45.4 | 45.7 | 45.9 | 46.3 | 46.6 | 47.0 | 47.2 | 47.2 |
| 26.0 | 42.1 | 42.7 | 43.3 | 43.7 | 44.0 | 44.3 | 44.7 | 45.0 | 45.3 | 45.5 | 45.5 |
| 27.0 | 40.5 | 41.1 | 41.7 | 42.1 | 42.4 | 42.7 | 43.0 | 43.5 | 43.8 | 43.9 | 43.8 |
| 28.0 | 38.9 | 39.6 | 40.1 | 40.5 | 40.8 | 41.1 | 41.5 | 41.9 | 42.2 | 42.3 | 42.3 |
| 29.0 | 37.4 | 38.0 | 38.5 | 38.9 | 39.3 | 39.6 | 40.0 | 40.4 | 40.7 | 40.9 | 40.8 |
| 30.0 | 35.8 | 36.5 | 37.0 | 37.4 | 37.8 | 38.1 | 38.5 | 38.9 | 39.3 | 39.4 | 39.4 |
| 32.0 | 33.3 | 33.9 | 34.3 | 34.8 | 35.1 | 35.5 | 35.8 | 36.3 | 36.6 | 36.7 | 36.7 |
| 34.0 | 30.9 | 31.4 | 31.8 | 32.2 | 32.6 | 33.0 | 33.3 | 33.8 | 34.0 | 34.1 | 34.1 |

From Palta, J.R., Ayyangar, K., Daftari, I., Suntharalingam, N., Characteristics of photon beams from Philips SL25 linear accelerators, *Med. Phys.*, 17, 106, 1990. With permission.

## TABLE 1.F8
## Tissue Phantom Ratios for 25-MV X Rays [Philips SL-25, 100 cm SAD]

| Depth (cm) | Field Size (cm×cm) | | | | | | | | | | |
|---|---|---|---|---|---|---|---|---|---|---|---|
| | 4×4 | 6×6 | 8×8 | 10×10 | 12×12 | 15×15 | 20×20 | 25×25 | 30×30 | 35×35 | 40×40 |
| 1.0 | 0.603 | 0.608 | 0.637 | 0.674 | 0.708 | 0.748 | 0.796 | 0.826 | 0.843 | 0.850 | 0.852 |
| 2.0 | 0.867 | 0.868 | 0.884 | 0.900 | 0.916 | 0.933 | 0.954 | 0.964 | 0.969 | 0.971 | 0.970 |
| 3.0 | 0.977 | 0.978 | 0.980 | 0.985 | 0.988 | 0.992 | 0.996 | 0.998 | 0.999 | 0.999 | 0.999 |
| 3.5 | 1.000 | 1.000 | 1.000 | 1.000 | 1.000 | 1.000 | 1.000 | 1.000 | 1.000 | 1.000 | 1.000 |
| 4.0 | 1.007 | 1.008 | 1.008 | 1.007 | 1.005 | 1.001 | 0.993 | 0.989 | 0.988 | 0.988 | 0.988 |
| 5.0 | 1.007 | 1.012 | 1.009 | 1.005 | 1.000 | 0.993 | 0.983 | 0.977 | 0.975 | 0.975 | 0.975 |
| 6.0 | 0.995 | 1.001 | 0.997 | 0.992 | 0.987 | 0.978 | 0.967 | 0.963 | 0.962 | 0.963 | 0.965 |
| 7.0 | 0.973 | 0.981 | 0.979 | 0.975 | 0.970 | 0.963 | 0.953 | 0.949 | 0.948 | 0.949 | 0.950 |
| 8.0 | 0.951 | 0.962 | 0.961 | 0.958 | 0.954 | 0.947 | 0.939 | 0.935 | 0.934 | 0.935 | 0.936 |
| 9.0 | 0.932 | 0.942 | 0.942 | 0.939 | 0.936 | 0.930 | 0.923 | 0.919 | 0.919 | 0.920 | 0.922 |
| 10.0 | 0.912 | 0.923 | 0.922 | 0.921 | 0.918 | 0.912 | 0.906 | 0.904 | 0.904 | 0.906 | 0.907 |
| 11.0 | 0.891 | 0.902 | 0.903 | 0.902 | 0.900 | 0.895 | 0.890 | 0.888 | 0.888 | 0.890 | 0.892 |
| 12.0 | 0.868 | 0.882 | 0.885 | 0.884 | 0.882 | 0.878 | 0.874 | 0.872 | 0.873 | 0.875 | 0.877 |
| 13.0 | 0.848 | 0.862 | 0.864 | 0.864 | 0.863 | 0.860 | 0.857 | 0.856 | 0.857 | 0.859 | 0.862 |
| 14.0 | 0.827 | 0.841 | 0.845 | 0.845 | 0.845 | 0.843 | 0.840 | 0.840 | 0.842 | 0.845 | 0.847 |
| 15.0 | 0.807 | 0.821 | 0.824 | 0.826 | 0.826 | 0.826 | 0.824 | 0.824 | 0.826 | 0.829 | 0.831 |
| 16.0 | 0.787 | 0.801 | 0.805 | 0.807 | 0.808 | 0.808 | 0.807 | 0.808 | 0.811 | 0.813 | 0.816 |
| 17.0 | 0.767 | 0.780 | 0.785 | 0.788 | 0.791 | 0.791 | 0.791 | 0.792 | 0.794 | 0.798 | 0.801 |
| 18.0 | 0.748 | 0.762 | 0.767 | 0.771 | 0.774 | 0.774 | 0.775 | 0.776 | 0.779 | 0.783 | 0.785 |
| 19.0 | 0.729 | 0.743 | 0.749 | 0.753 | 0.756 | 0.757 | 0.758 | 0.761 | 0.764 | 0.768 | 0.771 |
| 20.0 | 0.709 | 0.725 | 0.730 | 0.736 | 0.739 | 0.740 | 0.742 | 0.745 | 0.749 | 0.753 | 0.756 |
| 21.0 | 0.692 | 0.708 | 0.714 | 0.720 | 0.723 | 0.725 | 0.727 | 0.730 | 0.734 | 0.738 | 0.741 |
| 22.0 | 0.674 | 0.691 | 0.698 | 0.703 | 0.706 | 0.709 | 0.712 | 0.715 | 0.720 | 0.723 | 0.727 |
| 23.0 | 0.658 | 0.674 | 0.681 | 0.687 | 0.691 | 0.694 | 0.697 | 0.701 | 0.706 | 0.710 | 0.713 |
| 24.0 | 0.642 | 0.658 | 0.664 | 0.670 | 0.675 | 0.678 | 0.683 | 0.687 | 0.692 | 0.696 | 0.700 |
| 25.0 | 0.626 | 0.641 | 0.647 | 0.654 | 0.659 | 0.663 | 0.668 | 0.673 | 0.678 | 0.682 | 0.686 |
| 26.0 | 0.610 | 0.627 | 0.633 | 0.639 | 0.644 | 0.649 | 0.655 | 0.660 | 0.665 | 0.669 | 0.672 |
| 27.0 | 0.595 | 0.611 | 0.618 | 0.624 | 0.629 | 0.634 | 0.641 | 0.646 | 0.652 | 0.655 | 0.659 |
| 28.0 | 0.581 | 0.598 | 0.604 | 0.609 | 0.614 | 0.620 | 0.627 | 0.632 | 0.638 | 0.643 | 0.645 |
| 29.0 | 0.566 | 0.583 | 0.589 | 0.595 | 0.600 | 0.605 | 0.613 | 0.618 | 0.624 | 0.629 | 0.632 |
| 30.0 | 0.552 | 0.569 | 0.575 | 0.580 | 0.585 | 0.590 | 0.598 | 0.604 | 0.610 | 0.615 | 0.618 |
| 32.0 | 0.526 | 0.544 | 0.549 | 0.554 | 0.559 | 0.566 | 0.574 | 0.580 | 0.585 | 0.590 | 0.594 |
| 34.0 | 0.502 | 0.518 | 0.524 | 0.529 | 0.534 | 0.541 | 0.549 | 0.555 | 0.561 | 0.566 | 0.570 |

From Palta. J.R., Ayyangar, K., Daftari, I., Suntharalingam, N., Characteristics of photon beams from Philips SL25 linear accelerators, *Med. Phys.*, 17, 106, 1990. With permission.

## TABLE 1.F9
## Percent Depth Doses for 25-MV X Rays [AECL Therac-25, 100 cm SSD]

| Depth (cm) | Field Size (cm×cm) | | | | | | | | | | | | | |
|---|---|---|---|---|---|---|---|---|---|---|---|---|---|---|
| | 0×0 | 2×2 | 3×3 | 4×4 | 5×5 | 6×6 | 8×8 | 10×10 | 12×12 | 15×15 | 20×20 | 25×25 | 30×30 | 35×35 |
| 0.0 | 0.3 | 2.5 | 3.9 | 4.9 | 6.1 | 7.3 | 9.9 | 12.9 | 16.1 | 20.7 | 27.5 | 32.9 | 36.5 | 39.0 |
| 1.0 | 75.1 | 71.7 | 67.9 | 66.0 | 65.7 | 66.2 | 68.4 | 70.6 | 73.1 | 76.6 | 81.6 | 84.9 | 86.3 | 86.7 |
| 2.0 | 93.3 | 91.0 | 87.9 | 86.7 | 86.7 | 87.0 | 88.0 | 89.3 | 90.6 | 92.5 | 95.2 | 97.0 | 97.7 | 97.7 |
| 3.0 | 100.3 | 99.1 | 97.7 | 97.2 | 97.0 | 97.1 | 97.2 | 97.8 | 98.4 | 99.1 | 100.0 | 100.7 | 100.9 | 100.7 |
| 4.0 | 100.0 | 100.0 | 100.0 | 100.0 | 100.0 | 100.0 | 100.0 | 100.0 | 100.0 | 100.0 | 100.0 | 100.0 | 100.0 | 100.0 |
| 5.0 | 97.2 | 98.2 | 98.9 | 99.4 | 99.6 | 99.6 | 99.3 | 99.0 | 98.7 | 98.3 | 97.8 | 97.6 | 97.6 | 97.6 |
| 6.0 | 93.5 | 95.0 | 95.8 | 96.6 | 97.0 | 97.0 | 96.6 | 96.3 | 96.0 | 95.5 | 94.8 | 94.7 | 94.7 | 94.7 |
| 7.0 | 90.0 | 91.9 | 92.9 | 93.9 | 94.5 | 94.5 | 94.1 | 93.7 | 93.4 | 92.7 | 92.0 | 91.8 | 91.8 | 91.8 |
| 8.0 | 86.3 | 88.3 | 89.4 | 90.4 | 90.9 | 91.1 | 90.8 | 90.4 | 90.1 | 89.5 | 88.8 | 88.7 | 88.7 | 88.9 |
| 9.0 | 82.7 | 84.8 | 86.0 | 87.0 | 87.5 | 87.7 | 87.5 | 87.2 | 86.9 | 86.4 | 85.8 | 85.6 | 85.8 | 86.0 |
| 10.0 | 79.3 | 81.4 | 82.7 | 83.7 | 84.2 | 84.5 | 84.4 | 84.1 | 83.8 | 83.3 | 82.8 | 82.7 | 82.9 | 83.3 |
| 12.0 | 73.2 | 75.2 | 76.4 | 77.3 | 77.9 | 78.1 | 78.1 | 77.9 | 77.7 | 77.4 | 77.1 | 77.1 | 77.4 | 77.8 |
| 15.0 | 64.7 | 66.6 | 67.5 | 68.4 | 69.0 | 69.3 | 69.4 | 69.4 | 69.4 | 69.3 | 69.2 | 69.4 | 69.8 | 70.3 |
| 17.0 | 59.9 | 61.6 | 62.4 | 63.2 | 63.8 | 64.0 | 64.3 | 64.4 | 64.4 | 64.4 | 64.4 | 64.7 | 65.1 | 65.7 |
| 20.0 | 53.2 | 54.6 | 55.3 | 56.0 | 56.5 | 56.8 | 57.2 | 57.4 | 57.5 | 57.6 | 57.8 | 58.1 | 58.6 | 59.3 |
| 25.0 | 44.4 | 45.2 | 45.4 | 46.0 | 46.5 | 46.7 | 47.2 | 47.4 | 47.6 | 47.8 | 48.2 | 48.7 | 49.2 | 49.8 |
| 30.0 | 36.8 | 37.3 | 37.5 | 37.9 | 38.3 | 38.5 | 39.0 | 39.4 | 39.6 | 40.0 | 40.3 | 40.8 | 41.3 | 41.9 |

From Aldrich, J.E., Andrew, J.W., Michaels, H.B., and O'Brien, P.F., Characteristics of the photon beam from a new 25-MV linear accelerator, *Med. Phys.*, 12, 619, 1985. With permission.

## TABLE 1.F10
### Tissue Phantom Ratios for 25-MV X Rays [AECL Therac-25, 100 cm SAD]

| Depth (cm) | Field Size (cm×cm) 0×0 | 2×2 | 3×3 | 4×4 | 5×5 | 6×6 | 8×8 | 10×10 | 12×12 | 15×15 | 20×20 | 25×25 | 30×30 | 35×35 |
|---|---|---|---|---|---|---|---|---|---|---|---|---|---|---|
| 0.0 | 0.003 | 0.023 | 0.036 | 0.046 | 0.056 | 0.068 | 0.092 | 0.120 | 0.149 | 0.191 | 0.254 | 0.304 | 0.338 | 0.361 |
| 1.0 | 0.708 | 0.677 | 0.641 | 0.623 | 0.619 | 0.624 | 0.644 | 0.665 | 0.688 | 0.721 | 0.768 | 0.799 | 0.814 | 0.817 |
| 2.0 | 0.897 | 0.876 | 0.847 | 0.834 | 0.833 | 0.836 | 0.846 | 0.858 | 0.870 | 0.899 | 0.915 | 0.931 | 0.939 | 0.940 |
| 3.0 | 0.983 | 0.973 | 0.959 | 0.954 | 0.952 | 0.952 | 0.954 | 0.958 | 0.964 | 0.971 | 0.980 | 0.987 | 0.990 | 0.989 |
| 4.0 | 1.000 | 1.000 | 1.000 | 1.000 | 1.000 | 1.000 | 1.000 | 1.000 | 1.000 | 1.000 | 1.000 | 1.000 | 1.000 | 1.000 |
| 5.0 | 0.991 | 1.000 | 1.007 | 1.012 | 1.015 | 1.016 | 1.013 | 1.009 | 1.007 | 1.003 | 0.997 | 0.995 | 0.995 | 0.995 |
| 6.0 | 0.971 | 0.986 | 0.994 | 1.002 | 1.007 | 1.008 | 1.005 | 1.001 | 0.998 | 0.993 | 0.986 | 0.984 | 0.984 | 0.984 |
| 7.0 | 0.952 | 0.972 | 0.981 | 0.992 | 0.998 | 1.000 | 0.997 | 0.993 | 0.989 | 0.983 | 0.975 | 0.972 | 0.972 | 0.972 |
| 8.0 | 0.930 | 0.950 | 0.961 | 0.972 | 0.979 | 0.981 | 0.979 | 0.976 | 0.973 | 0.967 | 0.959 | 0.956 | 0.957 | 0.958 |
| 9.0 | 0.909 | 0.929 | 0.941 | 0.953 | 0.959 | 0.962 | 0.962 | 0.959 | 0.956 | 0.950 | 0.943 | 0.941 | 0.941 | 0.944 |
| 10.0 | 0.887 | 0.908 | 0.922 | 0.933 | 0.940 | 0.944 | 0.945 | 0.942 | 0.939 | 0.934 | 0.927 | 0.925 | 0.926 | 0.929 |
| 12.0 | 0.849 | 0.869 | 0.881 | 0.892 | 0.900 | 0.904 | 0.906 | 0.905 | 0.902 | 0.899 | 0.895 | 0.894 | 0.896 | 0.900 |
| 15.0 | 0.792 | 0.812 | 0.820 | 0.831 | 0.839 | 0.844 | 0.848 | 0.848 | 0.848 | 0.846 | 0.845 | 0.846 | 0.850 | 0.855 |
| 17.0 | 0.758 | 0.777 | 0.785 | 0.794 | 0.802 | 0.807 | 0.811 | 0.813 | 0.814 | 0.814 | 0.813 | 0.816 | 0.820 | 0.825 |
| 20.0 | 0.708 | 0.724 | 0.731 | 0.739 | 0.747 | 0.752 | 0.757 | 0.761 | 0.763 | 0.765 | 0.766 | 0.770 | 0.774 | 0.780 |
| 25.0 | 0.642 | 0.652 | 0.654 | 0.657 | 0.664 | 0.669 | 0.675 | 0.680 | 0.684 | 0.686 | 0.690 | 0.696 | 0.702 | 0.709 |
| 30.0 | 0.574 | 0.582 | 0.583 | 0.586 | 0.591 | 0.596 | 0.602 | 0.608 | 0.612 | 0.617 | 0.623 | 0.629 | 0.635 | 0.642 |

From Aldrich, J.E., Andrew, J.W., Michaels, H.B., and O'Brien, P.F., Characteristics of the photon beam from a new 25-MV linear accelerator, *Med. Phys.*, 12, 619, 1985. With permission.

## TABLE 1.G1
### Tissue Phantom Ratios for 33-MV X Rays [Brown-Boveri Betatron, 120 cm SAD]

| Depth (cm) | Field Size(cm×cm) | | | | | | | | |
|---|---|---|---|---|---|---|---|---|---|
| | 4×4 | 6×6 | 8×8 | 10×10 | 12×12 | 14×14 | 15×15 | 17×17 | 20×20 |
| 1.0 | 0.561 | 0.579 | 0.599 | 0.606 | 0.618 | 0.628 | 0.629 | 0.633 | 0.643 |
| 1.5 | 0.714 | 0.720 | 0.744 | 0.749 | 0.753 | 0.760 | 0.766 | 0.770 | 0.777 |
| 2.0 | 0.815 | 0.821 | 0.832 | 0.837 | 0.842 | 0.844 | 0.847 | 0.851 | 0.854 |
| 3.0 | 0.931 | 0.930 | 0.935 | 0.940 | 0.942 | 0.942 | 0.941 | 0.945 | 0.945 |
| 4.0 | 0.986 | 0.985 | 0.986 | 0.988 | 0.988 | 0.988 | 0.989 | 0.990 | 0.988 |
| 4.5 | 1.000 | 1.000 | 1.000 | 1.000 | 1.000 | 1.000 | 1.000 | 1.000 | 1.000 |
| 5.0 | 1.009 | 1.007 | 1.007 | 1.008 | 1.007 | 1.007 | 1.007 | 1.008 | 1.006 |
| 6.0 | 1.012 | 1.010 | 1.013 | 1.012 | 1.012 | 1.009 | 1.011 | 1.010 | 1.012 |
| 7.0 | 1.005 | 1.004 | 1.007 | 1.008 | 1.005 | 1.005 | 1.003 | 1.004 | 1.006 |
| 8.0 | 0.991 | 0.991 | 0.996 | 0.998 | 0.997 | 0.995 | 0.993 | 0.994 | 0.998 |
| 9.0 | 0.968 | 0.976 | 0.978 | 0.981 | 0.982 | 0.980 | 0.978 | 0.980 | 0.982 |
| 10.0 | 0.949 | 0.959 | 0.963 | 0.965 | 0.965 | 0.965 | 0.964 | 0.965 | 0.970 |
| 11.0 | 0.928 | 0.939 | 0.946 | 0.944 | 0.948 | 0.945 | 0.948 | 0.949 | 0.949 |
| 12.0 | 0.905 | 0.919 | 0.928 | 0.928 | 0.931 | 0.928 | 0.929 | 0.931 | 0.933 |
| 13.0 | 0.885 | 0.897 | 0.906 | 0.909 | 0.912 | 0.911 | 0.911 | 0.914 | 0.917 |
| 14.0 | 0.861 | 0.877 | 0.887 | 0.891 | 0.895 | 0.892 | 0.891 | 0.896 | 0.903 |
| 15.0 | 0.841 | 0.862 | 0.867 | 0.872 | 0.878 | 0.877 | 0.878 | 0.884 | 0.885 |
| 16.0 | 0.822 | 0.836 | 0.845 | 0.856 | 0.861 | 0.861 | 0.859 | 0.865 | 0.870 |
| 17.0 | 0.801 | 0.817 | 0.830 | 0.837 | 0.844 | 0.844 | 0.843 | 0.849 | 0.852 |
| 18.0 | 0.783 | 0.797 | 0.810 | 0.819 | 0.827 | 0.825 | 0.826 | 0.833 | 0.836 |
| 19.0 | 0.762 | 0.779 | 0.793 | 0.802 | 0.810 | 0.808 | 0.810 | 0.816 | 0.819 |
| 20.0 | 0.746 | 0.760 | 0.776 | 0.784 | 0.791 | 0.792 | 0.795 | 0.800 | 0.803 |
| 21.0 | 0.725 | 0.742 | 0.757 | 0.767 | 0.776 | 0.775 | 0.777 | 0.784 | 0.787 |
| 22.0 | 0.711 | 0.725 | 0.742 | 0.752 | 0.759 | 0.760 | 0.761 | 0.767 | 0.773 |
| 23.0 | 0.693 | 0.707 | 0.725 | 0.734 | 0.744 | 0.743 | 0.746 | 0.751 | 0.757 |
| 24.0 | 0.677 | 0.692 | 0.708 | 0.720 | 0.727 | 0.727 | 0.730 | 0.737 | 0.740 |
| 25.0 | 0.661 | 0.677 | 0.693 | 0.703 | 0.710 | 0.710 | 0.715 | 0.720 | 0.724 |
| 26.0 | 0.644 | 0.659 | 0.678 | 0.687 | 0.695 | 0.697 | 0.701 | 0.706 | 0.711 |
| 27.0 | 0.630 | 0.644 | 0.662 | 0.672 | 0.680 | 0.683 | 0.684 | 0.692 | 0.698 |

From Bagne, F., Physical aspects of supervoltage x-ray therapy, *Med. Phys.*, 1, 266, 1974. With permission.

**TABLE 1.G2**

**Percent Depth Doses for 34-MV X Rays [Asklepitron-35 Betatron, 110 cm SSD]**

| Depth (cm) | Field Size (cm×cm) | | | | | | | |
|---|---|---|---|---|---|---|---|---|
| | 5×5 | 6×6 | 7×7 | 8×8 | 10×10 | 12×12 | 15×15 | 20×20 |
| 0.0 | 12.2 | 13.4 | 15.4 | 17.2 | 20.6 | 22.9 | 25.4 | 28.4 |
| 0.5 | 45.2 | 46.1 | 47.2 | 48.6 | 51.2 | 53.3 | 55.4 | 56.9 |
| 1.0 | 66.0 | 66.8 | 67.6 | 68.4 | 70.1 | 71.7 | 73.4 | 74.3 |
| 1.5 | 78.3 | 78.5 | 78.9 | 79.4 | 80.7 | 82.4 | 84.1 | 84.4 |
| 2.0 | 86.4 | 86.5 | 86.7 | 87.0 | 88.3 | 89.6 | 90.5 | 90.8 |
| 2.5 | 92.1 | 92.1 | 92.1 | 92.4 | 93.4 | 94.2 | 95.0 | 95.0 |
| 3.0 | 95.5 | 95.5 | 95.7 | 96.0 | 96.6 | 97.1 | 97.6 | 97.6 |
| 3.5 | 97.8 | 97.8 | 97.8 | 97.8 | 98.5 | 98.9 | 98.9 | 98.9 |
| 4.0 | 99.2 | 99.3 | 99.3 | 99.5 | 99.8 | 100.0 | 100.0 | 100.0 |
| 4.5 | 100.0 | 100.0 | 100.0 | 100.0 | 100.0 | 100.0 | 100.0 | 100.0 |
| 5.0 | 100.0 | 100.0 | 100.0 | 100.0 | 100.0 | 100.0 | 100.0 | 99.4 |
| 5.5 | 100.0 | 100.0 | 100.0 | 100.0 | 99.4 | 99.2 | 99.0 | 98.4 |
| 6.0 | 99.6 | 99.5 | 99.5 | 99.3 | 98.5 | 98.2 | 97.8 | 97.6 |
| 6.5 | 98.8 | 98.7 | 98.6 | 98.5 | 97.6 | 97.4 | 96.7 | 96.4 |
| 7.0 | 97.4 | 97.3 | 97.2 | 97.0 | 96.3 | 95.8 | 95.5 | 95.3 |
| 8.0 | 94.7 | 94.7 | 94.7 | 94.5 | 93.9 | 93.6 | 92.7 | 92.3 |
| 9.0 | 91.5 | 91.9 | 92.0 | 91.9 | 91.2 | 90.5 | 89.6 | 88.8 |
| 10.0 | 88.6 | 88.7 | 88.7 | 88.6 | 87.8 | 86.9 | 86.4 | 86.3 |
| 11.0 | 85.5 | 85.5 | 85.4 | 85.2 | 84.6 | 83.9 | 83.3 | 83.2 |
| 12.0 | 82.4 | 82.4 | 82.2 | 82.0 | 81.4 | 80.8 | 80.2 | 80.2 |
| 13.0 | 79.5 | 79.3 | 79.1 | 79.0 | 78.5 | 78.0 | 77.3 | 77.3 |
| 14.0 | 76.4 | 76.3 | 76.1 | 75.9 | 75.4 | 75.0 | 74.4 | 74.4 |
| 15.0 | 73.4 | 73.4 | 73.2 | 73.0 | 72.5 | 72.2 | 71.7 | 71.7 |
| 16.0 | 70.4 | 70.5 | 70.4 | 70.1 | 69.6 | 69.5 | 69.4 | 69.3 |
| 17.0 | 67.8 | 67.9 | 67.8 | 67.6 | 67.0 | 66.9 | 66.9 | 66.9 |
| 18.0 | 65.2 | 65.3 | 65.3 | 65.2 | 64.7 | 64.6 | 64.6 | 64.6 |
| 19.0 | 62.6 | 62.7 | 62.8 | 62.7 | 62.3 | 62.3 | 62.3 | 62.3 |
| 20.0 | 60.2 | 60.4 | 60.5 | 60.4 | 60.2 | 60.2 | 60.2 | 60.2 |
| 22.0 | 56.0 | 56.2 | 56.1 | 56.0 | 56.0 | 56.0 | 56.0 | 56.0 |
| 24.0 | 52.2 | 52.2 | 52.2 | 52.2 | 52.2 | 52.2 | 52.2 | 52.3 |
| 26.0 | 48.1 | 48.3 | 48.4 | 48.4 | 48.5 | 48.5 | 48.6 | 48.8 |
| 28.0 | 44.6 | 44.8 | 44.9 | 45.0 | 45.0 | 45.1 | 45.2 | 45.4 |
| 30.0 | 41.3 | 41.5 | 41.6 | 41.7 | 41.8 | 41.9 | 42.1 | 42.4 |

From Dawson, D.J., Percentage depth doses for high energy x-rays, *Phys. Med. Biol.*, 21, 226, 1976. With permission.

## TABLE 1.G3
## Tissue Phantom Ratios for 45-MV X Rays [Brown-Boveri Betatron, 120 cm SAD]

| Depth (cm) | Field Size (cm×cm) | | | | | | |
|---|---|---|---|---|---|---|---|
| | 5×5 | 6×6 | 8×8 | 10×10 | 12×12 | 14×14 | 15×15 |
| 1.0 | 0.523 | 0.535 | 0.551 | 0.566 | 0.571 | 0.578 | 0.580 |
| 2.0 | 0.767 | 0.775 | 0.783 | 0.788 | 0.795 | 0.794 | 0.796 |
| 3.0 | 0.890 | 0.894 | 0.897 | 0.901 | 0.902 | 0.903 | 0.902 |
| 4.0 | 0.961 | 0.965 | 0.962 | 0.965 | 0.965 | 0.965 | 0.965 |
| 5.0 | 1.000 | 1.000 | 1.000 | 1.000 | 1.000 | 1.000 | 1.000 |
| 6.0 | 1.019 | 1.018 | 1.019 | 1.023 | 1.018 | 1.017 | 1.017 |
| 7.0 | 1.024 | 1.024 | 1.026 | 1.027 | 1.024 | 1.022 | 1.023 |
| 8.0 | 1.023 | 1.020 | 1.024 | 1.025 | 1.024 | 1.022 | 1.025 |
| 9.0 | 1.013 | 1.011 | 1.015 | 1.018 | 1.016 | 1.014 | 1.017 |
| 10.0 | 1.000 | 1.000 | 1.004 | 1.009 | 1.006 | 1.004 | 1.008 |
| 11.0 | 0.983 | 0.983 | 0.989 | 0.995 | 0.994 | 0.992 | 0.995 |
| 12.0 | 0.966 | 0.966 | 0.973 | 0.981 | 0.978 | 0.977 | 0.982 |
| 13.0 | 0.948 | 0.948 | 0.957 | 0.965 | 0.962 | 0.961 | 0.966 |
| 14.0 | 0.928 | 0.928 | 0.938 | 0.947 | 0.948 | 0.946 | 0.950 |
| 15.0 | 0.905 | 0.915 | 0.925 | 0.934 | 0.929 | 0.928 | 0.936 |
| 16.0 | 0.885 | 0.896 | 0.906 | 0.917 | 0.914 | 0.911 | 0.918 |
| 17.0 | 0.868 | 0.876 | 0.887 | 0.898 | 0.896 | 0.894 | 0.901 |
| 18.0 | 0.849 | 0.858 | 0.869 | 0.880 | 0.878 | 0.878 | 0.884 |
| 19.0 | 0.831 | 0.839 | 0.852 | 0.862 | 0.861 | 0.860 | 0.867 |
| 20.0 | 0.812 | 0.821 | 0.834 | 0.845 | 0.844 | 0.843 | 0.852 |
| 21.0 | 0.795 | 0.803 | 0.816 | 0.829 | 0.828 | 0.828 | 0.836 |
| 22.0 | 0.777 | 0.786 | 0.799 | 0.812 | 0.811 | 0.812 | 0.821 |
| 23.0 | 0.758 | 0.770 | 0.782 | 0.794 | 0.795 | 0.795 | 0.804 |
| 24.0 | 0.744 | 0.752 | 0.767 | 0.778 | 0.780 | 0.779 | 0.788 |
| 25.0 | 0.728 | 0.735 | 0.751 | 0.764 | 0.764 | 0.763 | 0.772 |
| 26.0 | 0.713 | 0.720 | 0.734 | 0.748 | 0.748 | 0.749 | 0.758 |
| 27.0 | 0.679 | 0.705 | 0.718 | 0.732 | 0.735 | 0.733 | 0.744 |

From Bagne, F., Physical aspects of supervoltage x-ray therapy, *Med. Phys.*, 1, 266, 1974. With permission.

**TABLE 1.H1**

**Surface (Skin) Dose Values for Various X Ray Beam Energies; SSD =100 cm**

| Field Size (cm × cm) | Co-60* | Relative Surface Dose (%) | | | | | | |
|---|---|---|---|---|---|---|---|---|
| | | 4-MV | 6-MV | 10-MV | 15-MV | 20-MV | 25-MV | 34-MV† |
| 5×5 | 13 | 13 | 8 | 6 | 7 | 8 | 6 | 12 |
| 10×10 | 21 | 18 | 13 | 12 | 11 | 13 | 13 | 21 |
| 15×15 | 30 | 23 | 18 | 18 | 16 | 17 | 21 | 25 |
| 20×20 | 38 | 30 | 23 | 23 | 22 | 22 | 27 | 28 |
| 30×30 | 55 | 38 | 31 | 32 | 29 | 28 | 36 | — |
| 40×40 | — | 43 | 36 | 35‡ | 33‡ | 32‡ | 41‡ | — |
| Ref. # | 4 | 69 | 69 | 68 | 39 | 34 | 40 | 51 |

* SSD = 80 cm.    † SSD = 110 cm.    ‡ Extrapolated values.

**TABLE 1.I1**

**Depth of Dose Maximum for Various X Ray Beam Energies; SSD = 100 cm**

| Field Size (cm×cm) | Co-60* | Depth of Dose Maximum, $d_{max}$ (cm) | | | | | | | | |
|---|---|---|---|---|---|---|---|---|---|---|
| | | 4-MV | 6-MV | 9-MV | 10-MV | 14-MV | 20-MV | 24-MV | 25-MV | 34-MV† |
| 5×5 | 0.5 | 1.00 | 1.6 | 2.0 | 2.4 | 2.9 | 3.8 | 4.1 | 4.0 | 5.0 |
| 10×10 | 0.5 | 1.00 | 1.5 | 1.9 | 2.3 | 2.7 | 3.5 | 3.9 | 3.8 | 4.7 |
| 15×15 | 0.5 | 1.00 | 1.5 | 1.8 | 2.1 | 2.5 | 3.2 | 3.5 | 3.4 | 4.5 |
| 20×20 | 0.5 | 0.90 | 1.4 | 1.7 | 1.9 | 2.4 | 2.7 | 3.0 | 2.8 | 4.3 |
| 30×30 | 0.5 | 0.90 | 1.4 | 1.6 | 1.8 | 2.3 | 2.6 | 2.9 | 2.6 | — |
| 40×40 | — | 0.85 | 1.3 | — | 1.8 | 2.2 | 2.4 | 2.8 | 2.4 | — |
| Ref. # | 4 | 10 | 22 | 24 | 32 | 36 | 34 | 42 | 44 | 51 |

* SSD = 80 cm.    † SSD = 110 cm.

**TABLE 1.J1**
**Summary of Central Axis Depth Dose Data for $10\times10$ cm$^2$ Photon Beams**

| Nominal Beam Energy | Source-to-Axis Distance (cm) | Depth of Dose Maximum $d_{max}$ (cm) | Relative Surface Dose | Percent Depth Dose at | | Tissue Phantom Ratio at | |
|---|---|---|---|---|---|---|---|
| | | | | 10 cm Depth | 20 cm Depth | 10 cm Depth | 20 cm Depth |
| Co-60 | 80.0 | 0.5 | 21.0 | 56.0 | 27.0 | 0.69 | 0.40 |
| 4-MV | 80.0 | 1.0 | 18.0 | 61.0 | 31.0 | 0.73 | 0.44 |
| 4-MV | 100.0 | 1.0 | 18.0 | 64.0 | 35.0 | 0.76 | 0.49 |
| 6-MV | 100.0 | 1.5 | 13.0 | 67.0 | 38.0 | 0.80 | 0.54 |
| 8-MV | 100.0 | 2.0 | 12.0 | 70.0 | 42.0 | 0.82 | 0.58 |
| 10-MV | 100.0 | 2.5 | 12.0 | 73.0 | 45.0 | 0.84 | 0.62 |
| 15-MV | 100.0 | 3.0 | 11.0 | 77.0 | 50.0 | 0.88 | 0.68 |
| 20-MV | 100.0 | 3.5 | 13.0 | 80.0 | 53.0 | 0.90 | 0.71 |
| 25-MV | 100.0 | 4.0 | 13.0 | 83.0 | 56.0 | 0.93 | 0.74 |
| 34-MV | 110.0 | 5.0 | 21.0 | 88.0 | 60.0 | 0.96 | 0.77 |

# REFERENCES

## A. COBALT-60 BEAMS

1.  **Chan, F.K., Haymond, H.R., Kagan, A.R., Carbone, G.E., and George III, F.W.**, Comparative beam data for the 25 MV Betatron, 8, 6, and 4 MV linear accelerators, and $^{60}$Co units, *Radiol.*, 109, 691, 1973.
2.  **Glasgow, G.**, Loyola University, Chicago, Illinois, personal communication, 1992.
3.  **Godden, T.J.**, Gamma radiation from Cobalt 60 teletherapy units, *Brit. J. Radiol.*, Suppl. 17, 37, 1983.
4.  The University of Iowa Hospitals and Clinics, Iowa City, IA, 1992.

## B. 3–4-MV X RAYS

5.  **Biggs, P.J., Doppke, K.P., Leong, J.C., and Russell, M.D.**, Tissue phantom ratios for a Clinac 4/100, *Med. Phys.*, 9, 753, 1982.
6.  **Castro, V.G. and Kopenhaver, J.F.**, Some aspects of dosimetry with a 4 MV linear accelerator, *Radiol.*, 102, 691, 1972.
7.  **Peterson, M. and Golden, R.**, Dosimetry of the Varian Clinac-4 linear accelerator, *Radiol.*, 103, 675, 1972.
8.  **Sable, M., Gunn, W.G., Penning, D., and Gardner, A.**, Performance of a new 4 MeV standing wave linear accelerator, *Radiol.*, 97, 169, 1970.
9.  **Steidley, K.D. and Rosen C.W.**, Dosimetric aspects of a 3.3-MV linear accelerator, *Med. Phys.*, 17, 474, 1990.
10. The University of Iowa Hospitals and Clinics, Iowa City, IA, 1992, personal communication.

## C. 6-MV X RAYS

11. **Al-Ghazi, M.S.A.L., Arjune, B., Fiedler, J.A., and Sharma, P.D.**, Dosimetric aspects of the therapeutic photon beams from a dual-energy linear accelerator, *Med. Phys.*, 15, 250, 1988.
12. **Coffey, II, C.W., Beach, J.L., Thompson, D.J., and Mendiondo, M.**, X-ray beam characteristics of the Varian Clinac 6-100 linear accelerator, *Med. Phys.*, 7, 716, 1980.
13. **Fontenla, D.P., Napoli, J.J., and Chui, C.S.**, Beam characteristics of a new model of 6-MV linear accelerator, *Med. Phys.*, 19, 343, 1992.
14. **George, R.E.**, Characteristics of an MM 22 medical microtron 6-MV photon beam, *Med. Phys.*, 11, 862, 1984.
15. **Grant III, W., Ames, J., and Almond, P.R.**, Evaluation of the Therac 6 linear accelerator for radiation therapy, *Med. Phys.*, 5, 448, 1978.
16. **Horsley, R.J., Price, R.H., Saunders, J.E., and Dingwall, P.W.**, Performance of a 6 MeV Varian linear accelerator, *Br. J. Radiol.*, 41, 312, 1968.
17. **Horton, J.L.**, Dosimetry of the Siemens mevatron 67 linear accelerator, *Int. J. Radiat. Oncol. Biol. Phys.*, 9, 1217, 1983.
18. **Ikoro, N.C., Johnson, D.A., and Antich, P.P.**, Characteristics of the 6-MV photon beam produced by a dual energy linear accelerator, *Med. Phys.*, 14, 93, 1987.
19. **Palta, J.R., Ayyangar, K., Daftari, I., Suntharalingam, N.**, Characteristics of photon beams from Philips SL25 linear accelerators, *Med. Phys.*, 17, 106, 1990.

20. **Palta, J.R., Ayyangar, K.M., and Suntharalingam, N.**, Dosimetric characteristics of a 6 MV photon beam from a linear accelerator with asymmetric collimator jaws, *Int. J. Radiat. Oncol. Biol. Phys.*, 14, 383, 1988.

21. **Sharma, S.C., Modur, P., and Basavatia, R.**, Evaluation of a photon and an electron beam of a 6-MV linear accelerator, *Med. Phys.*, 15, 525, 1988.

22. The University of Iowa Hospitals and Clinics, Iowa City, IA, 1992, personal communication.

## D. 8–10-MV X RAYS

23. **Agarwal, S.K., Scheele, R.V., and Wakley, J.**, Tissue maximum-dose ratio (TMR) for 8 MV x rays, *Am. J. Roentgenol. Rad. Ther. Nucl. Med.*, 112, 797, 1971.

24. **Arcovito, G., Piermattei, A., D'Abramo, G., and Bassi, F.A.**, Dose measurements and calculations of small radiation fields for 9-MV x rays, *Med. Phys.*, 12, 779, 1985.

25. **Chan, F.K., Haymond, H.R., Kagan, A.R., Carbone, G.E., and George III, F.W.**, Comparative beam data for the 25 MV Betatron, 8, 6, and 4 MV linear accelerators, and [60]Co units, *Radiol.*, 109, 691, 1973.

26. **Houdek, P.V., VanBuren, J.M., and Fayos, J.V.**, Dosimetry of small radiation fields for 10-MV x rays, *Med. Phys.*, 10, 333, 1983.

27. **Keller, B., Bassano, D., Mathewson, C., and Rubin, P.**, 10 MV photon beam characteristics: central axis depth doses, tissue-maximum ratios, scatter-maximum ratios, beam flatness, backscatter, and output factors, *Int. J. Radiat. Oncol. Biol. Phys.*, 1, 69, 1975.

28. **Khan, F.M.**, Depth dose and scatter analysis of 10 MV x-rays, *Radiol.*, 106, 662, 1973.

29. **Khan, F.M., Moore, V.C., and Sato, S.**, Depth dose and scatter analysis of 10 MV x-rays, *Radiol.*, 102, 165, 1972.

30. **Khan, F.M.**, *The Physics of Radiation Therapy*, Williams & Wilkins, Baltimore, 1984, 432.

31. **Szymczyk, W., Goraczko, A., and Lesiak, J.**, Prediction of Saturne II+ 10 MV and 23 MV photon beam output factors, *Int. J. Radiat. Oncol. Biol. Phys.*, 21, 789, 1991.

32. The University of Iowa Hospitals and Clinics, Iowa City, IA, 1992, personal communication.

33. **Wu, A., Leavitt, D.D., and Campbell, D.W.**, Dosimetry measurements and analyses of 10-MV x rays for Clinac-18, *Med. Phys.*, 4, 71, 1977.

## E. 14–20-MV X RAYS

34. **Al-Ghazi, M.S.A.L., Arjune, B., Fiedler, J.A., and Sharma, P.D.**, Dosimetric aspects of the therapeutic photon beams from a dual-energy linear accelerator, *Med. Phys.*, 15, 250, 1988.

35. **Johnson, D.A., Ikoro, N.C., Chang, C.-H., Scarbrough, E.C., and Antich, P.P.**, Properties of the 18-MV photon beam from a dual energy linear accelerator, *Med. Phys.*, 14, 1071, 1987.

36. **Mantel, J., Perry, H., and Weinkam, J.J.**, X-ray depth-dose characteristics of the Toshiba LMR-16, *Med. Phys.*, 6, 95, 1979.

37. **Palta, J.R., Meyer, J.A., and Hogstrom, K.R.**, Dosimetric characterization of the 18-MV photon beam from the Siemens Mevatron 77 linear accelerator, *Med. Phys.*, 11, 717, 1984.

38.  **Patterson, M.S. and Shragge, P.C.**, Characteristics of an 18 MV
     photon beam from a Therac 20 medical linear accelerator, *Med. Phys.*, 8,
     312, 1981.
39.  **Paul, J.M., Koch, R.F., Khan, F.R., and Devi, B.S.**, Characteristics of
     Mevatron 77 15-MV photon beam, *Med. Phys.*, 10, 237, 1983.

## F.  23–25-MV X RAYS
40.  **Aldrich, J.E., Andrew, J.W., Michaels, H.B., and O'Brien, P.F.**,
     Characteristics of the photon beam from a new 25-MV linear
     accelerator, *Med. Phys.*, 12, 619, 1985.
41.  **Chan, F.K., Haymond, H.R., Kagan, A.R., Carbone, G.E., and
     George III, F.W.**, Comparative beam data for the 25 MV betatron, 8, 6,
     and 4 MV linear accelerators, and $^{60}$Co units, *Radiol.*, 109, 691, 1973.
42.  **Jani, S.K. and Pennington, E.C.**, Depth dose characteristics of 24-MV
     x-ray beams at extended SSD, *Med. Phys.*, 18, 292, 1991.
43.  **Krithivas, G. and Rao, S.N.**, Dosimetry of 24 MV X rays from a linear
     accelerator, *Med. Phys.*, 14, 274, 1987.
44.  **Luxton, G. and Astrahan, M.A.**, Characteristics of the high-energy
     photon beam of a 25-MeV accelerator, *Med. Phys.*, 15, 82, 1988.
45.  **Palta, J.R., Ayyangar, K., Daftari, I., Suntharalingam, N.**,
     Characteristics of photon beams from Philips SL25 linear accelerators,
     *Med. Phys.*, 17, 106, 1990.
46.  **Purdy, J.A., Keys, D.J., and Abrath, F.G.**, 25 MV x-ray beam
     characteristics from a 35 MeV linear accelerator, *Int. J. Radiat. Oncol.
     Biol. Phys.*, 4, 337, 1978.
47.  **Szymczyk, W., Goraczko, A., and Lesiak, J.**, Prediction of Saturne II+
     10 MV and 23 MV photon beam output factors, *Int. J. Radiat. Oncol.
     Biol. Phys.*, 21, 789, 1991.
48.  The University of Iowa Hospitals and Clinics, Iowa City, IA, 1992,
     personal communication.

## G.  33–45-MV X RAYS
49.  **Bagne, F.**, Physical aspects of supervoltage x-ray therapy, *Med. Phys.*,
     1, 266, 1974.
50.  **Bhatnagar, J.P. and Spira, J.**, Variation of percentage depth dose with
     beam area of 43 MV Roentgen ray beam from a betatron, *Acta Radiol.
     Ther. Phys. Biol.*, 14, 337, 1975.
51.  **Dawson, D.J.**, Percentage depth doses for high energy x-rays, *Phys.
     Med. Biol.*, 21, 226, 1976.
52.  **Suntharalingam, N. and Steben, J.D.**, Physical characterization of 45-
     MV photon beams for use in treatment planning, *Med. Phys.*, 4, 134,
     1977.

## H, I.  SURFACE (SKIN) DOSE AND THE BUILDUP REGION
53.  **Ames, T.E., Saylor, W., and Dillard, M.**, Improvement of the skin
     sparing characteristics of the Clinac 4 by the use of leaded glass electron
     filters, *Int. J. Radiat. Oncol. Biol. Phys.*, 2, 1027, 1977.
54.  **Biggs, P.J. and Ling, C.C.**, Electrons as the cause of the observed
     $d_{max}$ shift with field size in high energy photon beams, *Med. Phys.*, 6,
     291, 1979.

55. **Bova, F.J. and Hill, L.W.,** Surface doses for acrylic versus lead acrylic blocking trays for Co-60, 8-MV, and 17-MV photons, *Med. Phys.*, 10, 254, 1983.

56. **Ciesielski, B., Reinstein, L.E., Wielopolski, L., and Meek, A.,** Dose enhancement in buildup region by lead, aluminum, and lucite absorbers for 15 MV photon beam, *Med. Phys.*, 16, 609, 1989.

57. **Gagnon, W.F. and Grant III, W.,** Surface dose from megavoltage therapy machines, *Radiol.*, 117, 705, 1975.

58. **Gagnon, W.F. and Horton, J.L.,** Physical factors affecting absorbed dose to the skin from cobalt-60 gamma rays and 25-MV x rays, *Med. Phys.*, 6, 285, 1979.

59. **Gagnon, W.F.,** Measurement of surface dose, *Int. J. Radiat. Oncol. Biol. Phys.*, 5, 449, 1979.

60. **Gerbi, B.J. and Khan, F.M.,** Measurement of dose in the buildup region using fixed-separation plane-parallel ionization chambers, *Med. Phys.*, 17, 17, 1990.

61. **Gerbi, B.J. and Khan, F.M.,** The polarity effect for commercially available plane-parallel ionization chambers, *Med. Phys.*, 14, 210, 1987.

62. **Khan, F.M., Moore, V.C., and Levitt, S.H.,** Effect of various atomic number absorbers on skin dose for 10-MeV x rays, *Radiol.*, 109, 209, 1973.

63. **Khan, F.M.,** Use of electron filter to reduce skin dose in cobalt teletherapy, *Amer. J. Roentgenol.*, 111, 180, 1971.

64. **Lambert, G.D., Liversage, W.E., Hirst, A.M., and Doughty, D.,** Exit dose studies in megavoltage photon therapy, *Brit. J. Radiol.*, 56, 329, 1983.

65. **Leung, P.M.K. and Johns, H.E.,** Use of electron filters to improve the buildup characteristics of large fields from Cobalt-60 beams, *Med. Phys.*, 4, 441, 1977.

66. **Leung, P.M.K., Sontag, M.R., Maharaj, H., and Chenery, S.,** Dose measurements in the build-up region for Cobalt-60 therapy units, *Med. Phys.*, 3, 169, 1976.

67. **Marbach, J.R. and Almond, P.R.,** Scattered photons as the cause for the observed $d_{max}$ shift with field size in high-energy photon beams, *Med. Phys.*, 4, 310, 1977.

68. **Mellenberg, Jr., D.E.,** Determination of build-up region over-response corrections for a Markus-type chamber, *Med. Phys.*, 17, 1041, 1990.

69. **Purdy, J.A.,** Buildup/surface dose and exit dose measurements for a 6-MV linear accelerator, *Med. Phys.*, 13, 259, 1986.

70. **Rubach, A., Conrad,, F., and Bichsel, H.,** Dose build-up curves for cobalt-60 irradiation: a systematic error occurring with pancake chamber measurements, *Phys. Med. Biol.*, 31, 441, 1986.

71. **Tannous, N.B.J., Gagnon, W.F., and Almond, P.R.,** Buildup region and skin-dose measurements for the Therac 6 linear accelerator for radiation therapy, *Med. Phys.*, 8, 378, 1981.

72. **Thomas, S.J. and Palmer, N.,** The use of carbon-loaded thermoluminescent dosimeters for the measurement of surface doses in megavoltage x-ray beams, *Med. Phys.*, 16, 902, 1989.

73. **Velkley, D.E., Manson, D.J., Purdy, J.A., and Oliver, Jr., G.D.,** Build-up region of megavoltage photon radiation sources, *Med. Phys.*, 2, 14, 1975.

74.  **Whitton, J.T.**, New values for epidermal thickness and their importance, *Health Phys.*, 24, 1, 1973.
75.  **Wu, A.**, Effects of an acrylic resin tray on relative surface doses for 10-MV x ray beams, *Int. J. Radiat. Oncol. Biol. Phys.*, 6, 1257, 1980.

## J. OTHER ARTICLES

76.  AAPM Task Group 21, A protocol for the determination of absorbed dose from high-energy photon and electron beams, *Med. Phys.*, 10, 741, 1983.
77.  **Day, M.J. and Aird, E.G.A.**, The equivalent field method for dose determinations in rectangular fields, *Brit. J. Radiol., Suppl.* 17, 105, 1983.
78.  **Gastorf, R.J., Hanson, W.F., Kirby, T.H., and Shalek, R.J.**, A comparison of high-energy accelerator depth dose data, *Med. Phys.*, 10, 881, 1983.
79.  **Greene, D. and Williams, P.C.**, X rays: 2–43 MV, *Brit. J. Radiol.*, Suppl. 17, 61, 1983.
80.  **Jani, S.K., Pennington, E.C., Wacha, J.E., and Anderson, K.M.**, Effect of collimator setting on the output of rectangular fields from linear accelerators, *Med. Dosimet.*, 13, 73, 1988.
81.  **Khan, F.M., Sewchand, W., Lee, J., and Williamson, J.F.**, Revision of tissue-maximum ratio and scatter-maximum ratio concepts for cobalt 60 and higher energy x-ray beams, *Med. Phys.*, 7, 230, 1980.
82.  **McCullough, E.C., Gortney, J., and Blackwell, C.R.**, A depth dependence determination of the wedge transmission factor for 4-10 MV photon beams, *Med. Phys.*, 15, 621, 1988.
83.  **Sterling, T.D., Perry, H., and Katz, L.**, Automation of radiation treatment planning-IV, *Brit. J. Radiol.*, 37, 544, 1964.
84.  **Van Dam, J., Bridier, A., Lasselin, C., Blanckaert, N., and Dutreix, A.**, Influence of shielding blocks on the output of photon beams as a function of energy and type of treatment unit, *Radiother. Oncol.*, 24, 55, 1992.

Chapter 2

# DATA ON ELECTRON BEAMS

The electron beam data are divided into three categories: depth dose data, surface dose and X ray contamination, and output for shaped electron beams. The dosimetric parameters associated with clinical electron beams are explained in the following paragraphs.

## A. DEPTH DOSE DATA

The central axis percent depth dose data for several megavoltage treatment machines are presented in Tables 2.A1 through 2.A14. In accordance with ICRU Report 35 recommendations,[22] the electron beam's central axis depth dose curve is characterized by several parameters, such as $R_{100}$, $R_{90}$, $R_{80}$, $R_{50}$, and $R_p$ as well as $D_s$ and $D_x$. These parameters are shown in Figure 2.1, and are defined as follows:

1. $R_{100}$ is the depth of dose maximum in water. Similarly, $R_{90}$, $R_{80}$ and $R_{50}$ represent the corresponding depths of 90%, 80% and 50% dose levels.
2. The practical range, $R_p$, determined from the depth dose or depth ionization curve, is the depth of the point where the tangent at the inflection point of the fall-off portion of the curve intersects the bremsstrahlung background. Practical range measurements using depth dose are only slightly different from those obtained from depth ionization curves.[22]
3. The therapeutic range $R_t$ is a measure of the clinically useful portion of the electron depth-dose profile. The AAPM Task Group 25[28] recommends the depth of the deepest 90% dose level (i.e., $R_{90}$) as the therapeutic range.
4. $D_s$ and $D_x$ are described in Section B.

The variation of depth-dose curve with energy is shown in Figure 2.2 for electron beams generated using a dual scattering foil mechanism. The surface dose, $D_s$, and the X ray dose $D_x$, increase with electron beam energy. Figure 2.3 shows isodose curves of an electron beam from a typical linear accelerator.

The depth dose parameters strongly depend upon electron energy that must be specified in a meaningful way. The electron beam that hits the accelerator window is nearly monoenergetic and has very little angular spread. As the beam passes through the exit window, scattering foil(s), monitor chamber, and other components of an accelerator, its energy spread is significantly increased. The most probable kinetic energy, $E_p$, of an electron beam at the phantom's surface, is related to the practical range, $R_p$, in water by the equation:

$$E_p = 0.22 + 1.98 * R_p + 0.0025 * R_p^2 \qquad (2.1)$$

where $E_p$ is in MeV and $R_p$ is in cm.[22, 46] The mean energy of the electron beam at the surface of the phantom, $E_o$, can be determined by the following relationship:

$$E_o = C * R_{50} \qquad (2.2)$$

where $C$ is a constant and $R_{50}$ is taken as the depth of either the 50% ionization or dose level. The value of $C$ has been controversial. The AAPM Task Group 21[28] protocol employed a value of 2.33 MeV/cm for $C$. Several other values for this constant have been reported in the literature.[26, 72, 80] For this book, we have employed the original AAPM Task Group 21 value of 2.33 MeV/cm for determining $E_o$ using the above equation 2.2. Thus, the values of $E_o$ listed in the third column of the electron depth dose data are either those obtained by the authors of the referred paper, or the ones we computed using $C = 2.33$ MeV/cm.

## B. SURFACE DOSE AND X RAY CONTAMINATION

1.  The surface (skin) dose $D_s$ is a dose measured at 0.5 mm depth relative to the maximum dose along the central ray. This definition of $D_s$ has been recommended by the AAPM Radiation Therapy Committee Task Group 25 on electron beam dosimetry.[28] The data presented here (Tables 2.B1 and 2.B2) are those reported in the literature or through private communication. Since most of the data were reported prior to the AAPM Task Group Report, they may represent $D_s$ as measured at "zero" depth. The reader is referred to the original reference for details on measurements.

2.  The X ray component of an electron beam dose, $D_x$, is due to bremsstrahlung interactions between the electrons and the accelerator end window, scattering foil, ionization (monitor) chambers, collimation system, and the patient. It is determined by extrapolating the bremsstrahlung tail to $R_p$. The dose $D_x$ increases with the energy of electron beam. A significant portion of $D_x$ seems to result from the scattering foil.[65] Therefore, the scanning electron beams with no scattering foil have lower $D_x$ values than the beams produced with a scattering foil mechanism.

## C. OUTPUT FOR SHAPED ELECTRON BEAMS

Clinically used, electron beams are often shaped to conform to the target volume. In modern accelerators, this field shaping (or blocking) is achieved at the end of the treatment cone, i.e., very close to the treatment SSD. The field shaping can alter the beam output (cGy/MU at depth of maximum dose) as well as the depth dose characteristics. For most clinical situations, the variation in output with field blocking is of primary concern, and should be accounted for in routine patient dosimetry. Several data sets from the literature have been compiled and are presented in Tables 2.C1 through 2.C7. Of secondary concern is the effect of blocking on beam penetration. Figure 2.4 shows how the depth dose characteristics change when extensive blocking is present.

The reference section contains several excellent papers on electron beam dosimetry. Often, the reader searches for a paper that deals either with a particular treatment machine's data, or the measurement techniques. A comprehensive listing of relevant published articles in this section will prove beneficial to the reader.

## LIST OF FIGURES

## LIST OF TABLES

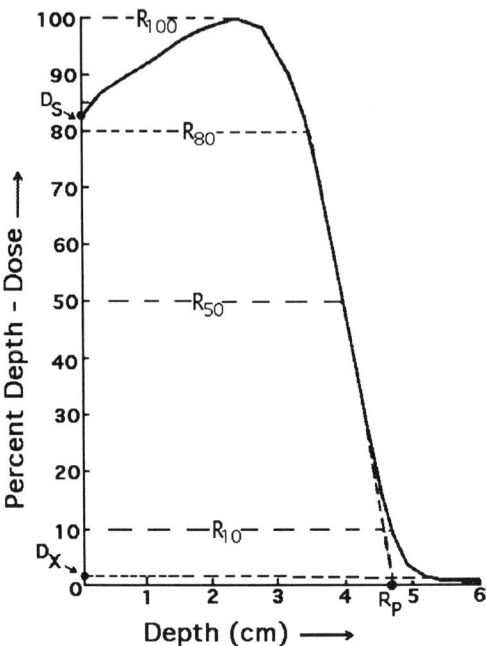

**FIGURE 2.1**: Central axis depth dose curve for an electron beam. Shown are the parameters used in characterizing electron beams.

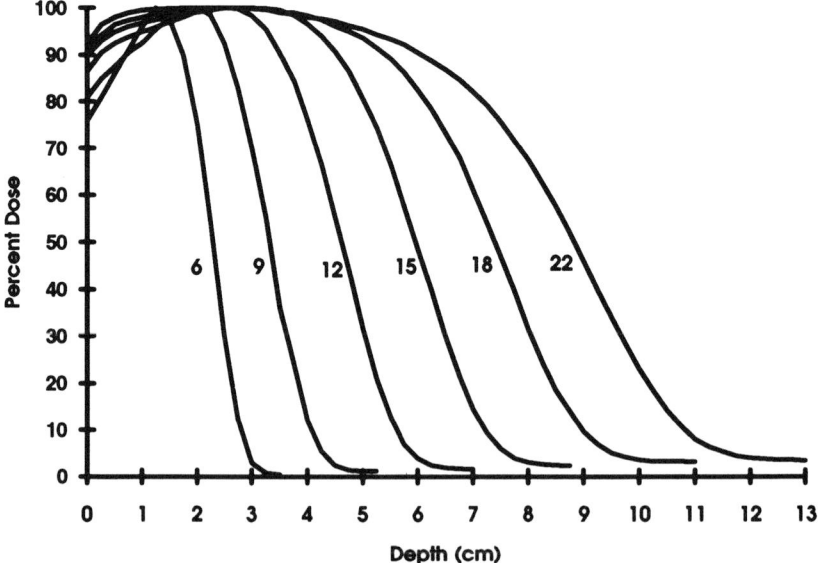

**FIGURE 2.2:** Central axis depth dose curves for electron beams of various energies. Varian Clinac-2500, 100 cm SSD, $15\times15$ cm$^2$ field size.

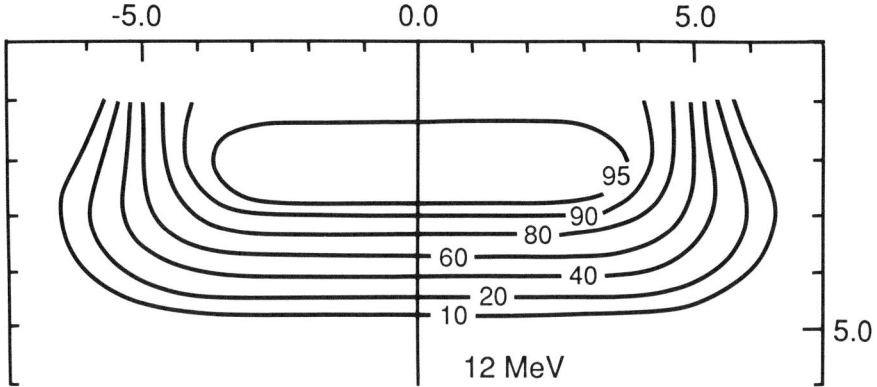

**FIGURE 2.3**: Isodose curves of a 12-MeV electron beam. Varian Clinac-2500, 100 cm SSD, 10×10 cm$^2$ field size.

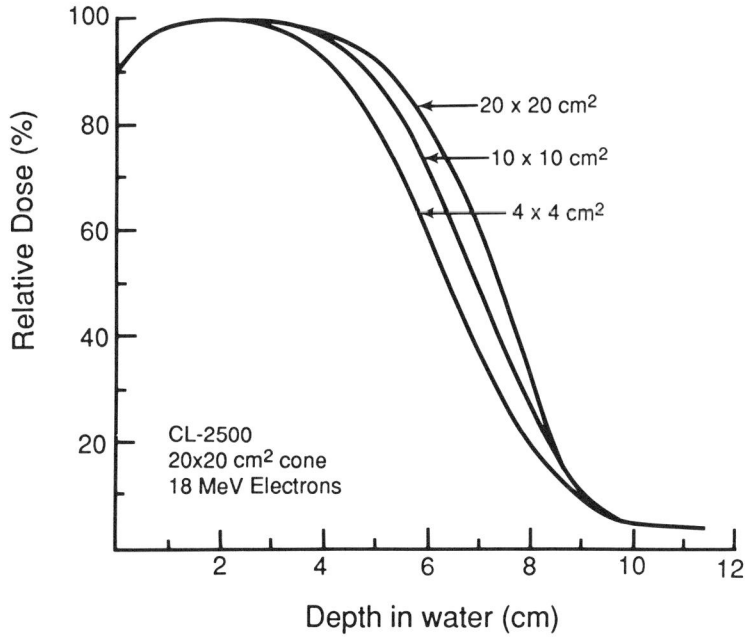

**FIGURE 2.4**: Effect of field shaping on depth dose curve of an 18-MeV electron beam. Varian Clinac-2500, 100 cm SSD.

## TABLE 2.A1
**Electron Beam Central Axis Depth Dose Data for Varian Clinac-18 [SSD=100 cm, Field size=15×15 cm²]**

| Nominal Energy $E_n$(MeV) | Most Probable Energy at Surface $E_p$ (MeV) | Mean Energy at Surface $E_o$(MeV) | Depth $R$ at which Dose $D$ Occurs $\{R_D \text{(cm)}\}$ | | | | | |
|---|---|---|---|---|---|---|---|---|
| | | | $R_{100}$ | $R_{90}$ | $R_{80}$ | $R_{50}$ | $R_P$ |
| 6 | 5.92 | 5.13 | 1.3 | 1.7 | 1.9 | 2.2 | 2.88 |
| 9 | 8.78 | 7.92 | 1.8 | 2.6 | 2.9 | 3.4 | 4.32 |
| 12 | 11.89 | 10.72 | 2.3 | 3.5 | 3.9 | 4.6 | 5.85 |
| 15 | 14.73 | 13.75 | 2.5 | 4.4 | 4.9 | 5.9 | 7.26 |
| 18 | 18.75 | 17.01 | 2.5 | 5.2 | 6.0 | 7.3 | 9.25 |

Adapted from Pennington, E.C. and Jani, S.K., The University of Iowa Hospitals and Clinics, Iowa City, Iowa, private communication, 1992.

## TABLE 2.A2
**Electron Beam Central Axis Depth Dose Data for Varian Clinac-20 [SSD=100 cm, Field Size=10×10 cm²]**

| Nominal Energy $E_n$(MeV) | Most Probable Energy at Surface $E_p$ (MeV) | Mean Energy at Surface $E_o$(MeV) | Depth $R$ at which Dose $D$ Occurs $\{R_D \text{(cm)}\}$ | | | | | |
|---|---|---|---|---|---|---|---|---|
| | | | $R_{100}$ | $R_{90}$ | $R_{80}$ | $R_{50}$ | $R_P$ |
| 6 | 5.98 | 5.4 | 1.0 | 1.6 | 1.9 | 2.3 | 2.9 |
| 9 | 9.18 | 8.2 | 1.6 | 2.6 | 2.9 | 3.5 | 4.5 |
| 12 | 11.39 | 10.5 | 2.2 | 3.4 | 3.8 | 4.5 | 5.6 |
| 16 | — | 14.2 | — | — | 5.4 | 6.1 | — |
| 20 | — | 19.1 | — | — | 6.4 | 8.2 | — |

From Kirby, T.H., Gastorf, R.J., Hanson, W.F., Berkley, L.W., Gagnon, W.F., Hazle, J.D., and Shalek, R.J., Electron beam central axis depth dose measurements, *Med. Phys.*, 12, 357, 1985. With permission.

**TABLE 2.A3**

**Electron Beam Central Axis Depth Dose Data for Varian Clinac-2500 [SSD=100 cm, Field Size=15×15 cm²]**

| Nominal Energy $E_n$(MeV) | Most Probable Energy at Surface $E_p$(MeV) | Mean Energy at Surface $E_o$(MeV) | Depth $R$ at which Dose $D$ Occurs $\{R_D\text{(cm)}\}$ | | | | |
|---|---|---|---|---|---|---|---|
| | | | $R_{100}$ | $R_{90}$ | $R_{80}$ | $R_{50}$ | $R_P$ |
| 6 | 5.72 | 5.13 | 1.3 | 1.6 | 1.8 | 2.2 | 2.77 |
| 9 | 8.42 | 7.69 | 1.9 | 2.6 | 2.8 | 3.3 | 4.12 |
| 12 | 11.69 | 10.72 | 2.5 | 3.5 | 3.9 | 4.6 | 5.75 |
| 15 | 14.93 | 13.75 | 2.7 | 4.4 | 5.0 | 5.9 | 7.36 |
| 18 | 18.55 | 17.24 | 2.7 | 5.5 | 6.2 | 7.4 | 9.15 |
| 22 | 22.10 | 20.50 | 2.0 | 6.1 | 7.2 | 8.8 | 10.90 |

Adapted from Pennington, E.C. and Jani, S.K., The University of Iowa Hospitals and Clinics, Iowa City, Iowa, private communication, 1992.

**TABLE 2.A4**

**Electron Beam Central Axis Depth Dose Data for Siemens Mevatron XII [SSD=100 cm, Field Size=10×10 cm²]**

| Nominal Energy $E_n$(MeV) | Most Probable Energy at Surface $E_p$(MeV) | Mean Energy at Surface $E_o$(MeV) | Depth $R$ at which Dose $D$ Occurs $\{R_D\text{(cm)}\}$ | | | | |
|---|---|---|---|---|---|---|---|
| | | | $R_{100}$ | $R_{90}$ | $R_{80}$ | $R_{50}$ | $R_P$ |
| 7 | — | 6.1 | 1.1 | 2.2 | 2.3 | 2.6 | — |
| 8 | 7.8 | 7.0 | 1.5 | 2.2 | 2.5 | 3.0 | 3.8 |
| 11 | 10.6 | 9.6 | 2.2 | 3.1 | 3.5 | 4.1 | 5.2 |
| 79 | 11.8 | 11.0 | 2.8 | 3.8 | 4.1 | 4.7 | 5.8 |

From Kirby, T.H., Gastorf, R.J., Hanson, W.F., Berkley, L.W., Gagnon, W.F., Hazle, J.D., and Shalek, R.J., Electron beam central axis depth dose measurements, *Med. Phys.*, 12, 357, 1985. With permission.

## TABLE 2.A5
### Electron Beam Central Axis Depth Dose Data for Siemens Mevatron XX [SSD=100 cm, Field Size=10×10 cm$^2$]

| Nominal Energy $E_n$(MeV) | Most Probable Energy at Surface $E_p$(MeV) | Mean Energy at Surface $E_o$(MeV) | Depth $R$ at which Dose $D$ Occurs $\{R_D$(cm)$\}$ | | | | | |
|---|---|---|---|---|---|---|---|---|
| | | | $R_{100}$ | $R_{90}$ | $R_{80}$ | $R_{50}$ | $R_P$ |
| 5 | 4.6 | 4.2 | 1.0 | 1.3 | 1.5 | 1.8 | 2.2 |
| 7 | 7.0 | 6.3 | 1.6 | 2.0 | 2.2 | 2.7 | 3.4 |
| 10 | 10.4 | 9.7 | 2.4 | 3.2 | 3.5 | 4.1 | 5.1 |
| 12 | 12.6 | 11.9 | 2.9 | 4.0 | 4.4 | 5.1 | 6.2 |
| 15 | 15.6 | 15.2 | 3.0 | 5.0 | 5.5 | 6.5 | 7.7 |
| 18 | 18.4 | 17.5 | 3.0 | 5.7 | 6.3 | 7.6 | 9.1 |

From Niroomand-Rad, A., Gillin, M.T., Kline, R.W., and Grimm, D.F., Film dosimetry of small electron beams for routine radiotherapy planning, *Med. Phys.*, 13, 416, 1986. With permission.

## TABLE 2.A6
### Electron Beam Central Axis Depth Dose Data for Siemens Mevatron-77 [SSD=100 cm, Field Size=10×10 cm$^2$]

| Nominal Energy $E_n$(MeV) | Most Probable Energy at Surface $E_p$(MeV) | Mean Energy at Surface $E_o$(MeV) | Depth $R$ at which Dose $D$ Occurs $\{R_D$(cm)$\}$ | | | | | |
|---|---|---|---|---|---|---|---|---|
| | | | $R_{100}$ | $R_{90}$ | $R_{80}$ | $R_{50}$ | $R_P$ |
| 7 | 7.0 | 6.2 | 1.5 | 2.0 | 2.2 | 2.65 | 3.4 |
| 12 | 11.8 | 10.7 | 2.6 | 3.6 | 3.9 | 4.6 | 5.8 |
| 18 | 17.8 | 16.1 | 2.6 | 5.0 | 5.7 | 6.9 | 8.8 |

From Shiu, A.S., Otte, V.A., and Hogstrom K.R., Measurement of dose distributions using film in therapeutic electron beams, *Med. Phys.*, 16, 911, 1989. With permission.

## TABLE 2.A7
### Electron Beam Central Axis Depth Dose Data for Siemens Mevatron-80 [SSD=100 cm, Field Size=10×10 cm$^2$]

| Nominal Energy $E_n$(MeV) | Most Probable Energy at Surface $E_p$(MeV) | Mean Energy at Surface $E_0$(MeV) | Depth $R$ at which Dose $D$ Occurs {$R_D$(cm)} | | | | |
|---|---|---|---|---|---|---|---|
| | | | $R_{100}$ | $R_{90}$ | $R_{80}$ | $R_{50}$ | $R_P$ |
| 7 | 6.8 | 6.3 | 1.6 | 2.1 | 2.3 | 2.7 | 3.3 |
| 10 | 9.8 | 9.3 | 2.4 | 3.1 | 3.4 | 4.0 | 4.8 |
| 12 | 11.8 | 11.1 | 2.8 | 3.7 | 4.1 | 4.75 | 5.8 |
| 15 | 14.8 | 13.9 | 2.6 | 4.5 | 5.0 | 5.95 | 7.3 |
| 18 | 18.0 | 16.9 | 1.6 | 5.3 | 6.0 | 7.25 | 8.9 |

From Meyer, J.A., Palta, J.R., and Hogstrom K.R., Demonstration of relatively new electron dosimetry measurement techniques on the Mevatron 80, *Med. Phys.*, 11, 670, 1984. With permission.

## TABLE 2.A8
### Electron Beam Central Axis Depth Dose Data for Philips SL-20 [SSD=100 cm, Field Size=10×10 cm$^2$]

| Nominal Energy $E_n$(MeV) | Most Probable Energy at Surface $E_p$(MeV) | Mean Energy at Surface $E_0$(MeV) | Depth $R$ at which Dose $D$ Occurs {$R_D$(cm)} | | | | |
|---|---|---|---|---|---|---|---|
| | | | $R_{100}$ | $R_{90}$ | $R_{80}$ | $R_{50}$ | $R_P$ |
| 4 | 4.4 | 3.7 | 0.8 | 1.1 | 1.3 | 1.6 | 2.1 |
| 6 | 6.4 | 5.6 | 1.4 | 1.9 | 2.0 | 2.4 | 3.1 |
| 8 | 8.0 | 7.5 | 1.8 | 2.4 | 2.7 | 3.2 | 3.9 |
| 10 | 9.8 | 9.1 | 2.0 | 3.0 | 3.3 | 3.9 | 4.8 |
| 12 | 11.6 | 10.9 | 2.5 | 3.6 | 3.9 | 4.7 | 5.7 |
| 15 | 14.2 | 13.3 | 2.5 | 4.3 | 4.8 | 5.7 | 7.0 |
| 18 | 18.0 | 16.1 | 2.5 | 5.1 | 5.7 | 6.6 | 8.9 |
| 20 | 19.7 | 18.4 | 2.5 | 5.9 | 6.6 | 7.9 | 9.7 |

From Lamba, M.A.S. and Elson, H.R., The University of Cincinnati Medical Center, Cincinnati, Ohio, private communication, 1992. With permission.

## TABLE 2.A9
### Electron Beam Central Axis Depth Dose Data for Philips SL-25 [SSD=95 cm, Field Size=15×15 cm²]

| Nominal Energy $E_n$(MeV) | Most Probable Energy at Surface $E_p$ (MeV) | Mean Energy at Surface $E_0$(MeV) | Depth $R$ at which Dose $D$ Occurs {$R_D$(cm)} | | | | | |
|---|---|---|---|---|---|---|---|---|
| | | | $R_{100}$ | $R_{90}$ | $R_{80}$ | $R_{50}$ | $R_P$ | |
| 4 | 4.6 | 3.8 | 1.2 | 1.25 | 1.65 | 2.2 | 2.2 |
| 6 | 6.4 | 5.7 | 1.9 | 1.95 | 2.47 | 3.1 | 3.1 |
| 8 | 8.2 | 7.5 | 2.45 | 2.6 | 3.22 | 4.0 | 4.0 |
| 10 | 9.6 | 9.0 | 3.0 | 3.15 | 3.86 | 4.7 | 4.7 |
| 12 | 11.4 | 10.7 | 3.6 | 3.75 | 4.59 | 5.5 | 5.6 |
| 15 | 14.4 | 13.5 | 4.5 | 4.75 | 5.79 | 7.0 | 7.1 |
| 17 | 15.8 | 15.0 | 4.95 | 5.2 | 6.44 | 7.7 | 7.8 |
| 20 | 19.4 | 18.4 | 5.9 | 6.3 | 7.90 | 7.8 | 9.5 |
| 22 | 21.4 | 20.2 | 6.2 | 6.8 | 8.67 | 10.6 | 10.5 |

From Palta, J.R., Daftari, I.K., Ayyanger, K.M., and Suntharalingam, N., Electron beam characteristics on a Philips SL25, *Med. Phys.*, 17, 27, 1990. With permission.

**TABLE 2.A10**
**Electron Beam Central Axis Depth Dose Data for Scanditronix MM-22 [SSD=100 cm, Field Size=10×10 cm$^2$]**

| Nominal Energy $E_n$(MeV) | Most Probable Energy at Surface $E_p$(MeV) | Mean Energy at Surface $E_o$(MeV) | Depth $R$ at which Dose $D$ Occurs {$R_D$(cm)} | | | | |
|---|---|---|---|---|---|---|---|
| | | | $R_{100}$ | $R_{90}$ | $R_{80}$ | $R_{50}$ | $R_P$ |
| 3 | 3.4 | 2.6 | 0.5 | 0.8 | 0.9 | 1.1 | 1.6 |
| 5 | 5.0 | 4.7 | 1.0 | 1.45 | 1.65 | 2.0 | 2.4 |
| 7 | 7.6 | 7.0 | 1.6 | 2.3 | 2.5 | 3.0 | 3.7 |
| 9 | 9.8 | 9.1 | 2.2 | 3.0 | 3.3 | 3.9 | 4.8 |
| 11 | 12.0 | 11.2 | 2.5 | 3.7 | 4.1 | 4.8 | 5.9 |
| 13 | 13.9 | 13.3 | 3.0 | 4.4 | 4.8 | 5.7 | 6.8 |
| 16 | 16.2 | 15.4 | 3.0 | 5.5 | 6.1 | 6.6 | 8.0 |
| 18 | 18.3 | 17.5 | 3.0 | 6.1 | 6.8 | 7.5 | 9.0 |
| 20 | 20.9 | 20.0 | 3.0 | 7.1 | 7.9 | 8.6 | 10.3 |
| 22 | 22.5 | 21.4 | 3.0 | 7.6 | 8.5 | 9.2 | 11.1 |

From George, R.E., Frost, S.V., and Hartson-Eaton, M., Characteristics of electron beams from a medical microtron, *Med. Phys.*, 13, 533, 1986. With permission.

# TABLE 2.A11
## Electron Beam Central Axis Depth Dose Data for AECL Therac-20 [SSD=100 cm, *Field Size=25×25 cm²]

| Nominal Energy $E_n$(MeV) | Most Probable Energy at Surface $E_p$(MeV) | Mean Energy at Surface $E_0$(MeV) | Depth $R$ at which Dose $D$ Occurs {$R_D$(cm)} | | | | | |
|---|---|---|---|---|---|---|---|---|
| | | | $R_{100}$ | $R_{90}$ | $R_{80}$ | $R_{50}$ | $R_P$ | |
| 6 | 5.4 | 4.7 | 1.2 | 1.6 | 1.7 | 2.0 | 2.6 | |
| 9 | 8.4 | 7.9 | 2.0 | 2.6 | 2.8 | 3.5 | 4.1 | |
| 13 | 12.4 | 11.6 | 2.9 | 3.8 | 4.2 | 4.9 | 6.1 | |
| 17 | 16.4 | 15.7 | 3.6 | 4.9 | 5.5 | 6.6 | 8.1 | |
| 20 | 19.5 | 18.5 | 4.0 | 5.7 | 6.3 | 7.8 | 9.6 | |

* Field size = 10×10 cm² for 6 MeV; 15×15 cm² for 9 MeV.

From Chen, F-S., An analytical equation of electron beams percentage depth ionization curve along the central axis, *Med. Phys.*, 15, 407, 1988. With permission.

## TABLE 2.A12

### Electron Beam Central Axis Depth Dose Data for AECL Therac-25 [SSD=100 cm, Field Size=10×10 cm²]

| Nominal Energy $E_n$(MeV) | Most Probable Energy at Surface $E_p$(MeV) | Mean Energy at Surface $E_o$(MeV) | Depth $R$ at which Dose $D$ Occurs {$R_D$(cm)} | | | | |
|---|---|---|---|---|---|---|---|
| | | | $R_{100}$ | $R_{90}$ | $R_{80}$ | $R_{50}$ | $R_P$ |
| 5 | 4.0 | 3.3 | 0.7 | 1.1 | 1.2 | 1.4 | 1.9 |
| 7 | 6.2 | 5.6 | 1.4 | 1.9 | 2.0 | 2.4 | 3.0 |
| 10 | 8.8 | 7.9 | 2.0 | 2.7 | 2.9 | 3.4 | 4.3 |
| 13 | 11.8 | 11.0 | 2.8 | 3.9 | 4.2 | 4.7 | 5.8 |
| 16 | 15.6 | 14.7 | 3.5 | 5.1 | 5.5 | 6.3 | 7.7 |
| 19 | 18.0 | 17.0 | 4.3 | 5.9 | 6.5 | 7.3 | 8.9 |
| 22 | 21.9 | 20.7 | 4.4 | 7.2 | 7.9 | 8.9 | 10.8 |
| 25 | 24.7 | 23.3 | 4.0 | 7.6 | 8.7 | 10.0 | 12.2 |

From O'Brien, P., Michaels, H.B., Aldrich J.E., and Andrew, J.W., Characteristics of electron beams from a new 25-MeV linear accelerator, *Med. Phys.*, 12, 799, 1985. With permission.

## TABLE 2.A13

### Electron Beam Central Axis Depth Dose Data for CGR Sagittaire [SSD=100 cm, Field Size=10×10 cm²]

| Nominal Energy $E_n$(MeV) | Most Probable Energy at Surface $E_p$(MeV) | Mean Energy at Surface $E_o$(MeV) | Depth $R$ at which Dose $D$ Occurs {$R_D$(cm)} | | | | |
|---|---|---|---|---|---|---|---|
| | | | $R_{100}$ | $R_{90}$ | $R_{80}$ | $R_{50}$ | $R_P$ |
| 10 | 10.0 | 9.3 | 2.4 | 3.2 | 3.5 | 4.0 | 4.9 |
| 13 | 13.0 | 12.3 | 3.0 | 4.1 | 4.5 | 5.3 | 6.4 |
| 16 | 14.4 | 14.2 | 3.8 | 4.8 | 5.2 | 6.1 | 7.1 |
| 22 | 22.1 | 21.4 | 3.8 | 7.0 | 7.8 | 9.2 | 10.9 |

From Kirby, T.H., Gastorf, R.J., Hanson, W.F., Berkley, L.W., Gagnon, W.F., Hazle, J.D., and Shalek, R.J., Electron beam central axis depth dose measurements, *Med. Phys.*, 12, 357, 1985. With permission.

## TABLE 2.A14
### Electron Beam Central Axis Depth Dose Data for Allis-Chalmer Betatron [SSD=100 cm, Field Size=$10 \times 10$ cm$^2$]

| Nominal Energy $E_n$(MeV) | Most Probable Energy at Surface $E_p$ (MeV) | Mean Energy at Surface $E_o$(MeV) | Depth R at which Dose D Occurs {$R_D$ (cm)} | | | | |
|---|---|---|---|---|---|---|---|
| | | | $R_{100}$ | $R_{90}$ | $R_{80}$ | $R_{50}$ | $R_P$ |
| 9 | 8.8 | 7.5 | 1.8 | 2.4 | 2.6 | 3.2 | 4.3 |
| 10 | 9.2 | 8.4 | 1.7 | 2.6 | 2.9 | 3.6 | 4.5 |
| 12 | 12.2 | 10.7 | 1.3 | 2.9 | 3.4 | 4.6 | 6.0 |
| 15 | 13.4 | 11.9 | 1.5 | 3.3 | 4.0 | 5.1 | 6.6 |
| 10* | 10.8 | 10.0 | 2.1 | 3.3 | 3.6 | 4.3 | 5.3 |
| 12** | 11.8 | 11.0 | 2.5 | 3.6 | 4.0 | 4.7 | 5.8 |

* Brown-Boveri Betatron

** Siemens Betatron

From Kirby, T.H., Gastorf, R.J., Hanson, W.F., Berkley, L.W., Gagnon, W.F., Hazle, J.D., and Shalek, R.J., Electron beam central axis depth dose measurements, *Med. Phys.*, 12, 357, 1985. With permission.

**TABLE 2.B1**
**Surface (Skin) Dose for Electron Beams**

| Nominal Beam Energy (MeV) | Surface Dose $D_s$ (% of $D_{max}$) for 10×10 to 15×15 cm² field | | | | | |
| --- | --- | --- | --- | --- | --- | --- |
| | Varian CL 2500 (Ref. #53) | Varian CL 18 (Ref. #53) | Siemens Mevatron XX (Ref. #49) | Philips SL 25 (Ref. #52) | Scanditronix MM 22 (Ref. #16) | AECL Therac 25 (Ref. #50) |
| 3 | — | — | — | — | 75.0 | — |
| 4 | — | — | — | 80.0 | 72.0 | — |
| 5 | 78.0 | — | 76.5 | — | — | 71.0 |
| 6 | — | 78.0 | — | 82.0 | 77.0 | — |
| 7 | — | — | 79.5 | — | — | 75.0 |
| 8 | — | 83.0 | — | 84.0 | 81.0 | — |
| 9 | 81.0 | — | — | — | — | — |
| 10 | — | — | 86.6 | 85.0 | 84.0 | 81.0 |
| 11 | 86.0 | 86.0 | — | — | — | — |
| 12 | — | — | 89.6 | 88.0 | 88.0 | 84.0 |
| 13 | 89.0 | 90.0 | — | — | — | — |
| 15 | — | — | 93.0 | 91.0 | 90.0 | 89.0 |
| 16 | — | — | — | — | — | — |
| 17 | 91.0 | 91.0 | — | 91.0 | 92.0 | — |
| 18 | — | — | 93.7 | — | — | — |
| 19 | — | — | — | — | — | 92.0 |
| 20 | — | — | — | 92.0 | 94.0 | — |
| 22 | 92.5 | — | — | 92.0 | 96.0 | 94.0 |
| 25 | — | — | — | — | — | 95.0 |

**TABLE 2.B2**
**X Ray Background Dose ($D_x$) for Electron Beams**

| Nominal Beam Energy (MeV) | Varian CL 2500 (Ref. #53) | Varian CL 18 (Ref. #53) | Siemens Mevatron XX (Ref. #49, 50) | Philips SL 25 (Ref. #52) | Scanditronix MM 22 (Ref. #16) | AECL Therac 25 (Ref. #50) |
|---|---|---|---|---|---|---|
| | | | X Ray Dose $D_x$ (% of $D_{max}$) for 10×10 to 15×15 cm² field | | | |
| 3 | — | — | — | — | 1.0 | — |
| 4 | — | — | — | <1.0 | 0.6 | — |
| 5 | 0.5 | — | <0.5 | — | — | <0.5 |
| 6 | — | 0.6 | — | <1.0 | — | — |
| 7 | — | — | 0.5 | 1.0 | 1.5 | <0.5 |
| 8 | — | — | — | 1.0 | 1.5 | — |
| 9 | 1.5 | 2.5 | — | 1.3 | — | — |
| 10 | — | — | 1.3 | — | — | 0.6 |
| 11 | — | — | — | — | 1.7 | — |
| 12 | 1.7 | 3.0 | 2.2 | 1.8 | — | — |
| 13 | — | — | — | — | 3.1 | 1.0 |
| 15 | 2.5 | 4.0 | 3.6 | 1.8 | — | — |
| 16 | — | — | — | — | 3.7 | 1.6 |
| 17 | — | — | — | 2.2 | — | — |
| 18 | 2.5 | 4.5 | 4.9 | — | 4.1 | — |
| 19 | — | — | — | — | — | 2.0 |
| 20 | — | — | — | 3.1 | 3.3 | — |
| 22 | 3.5 | — | — | 3.8 | 3.4 | 2.8 |
| 25 | — | — | — | — | — | 3.1 |

## TABLE 2.C1
## Effect of Field Shaping on Electron Beam Output for Varian Clinac-18, SSD=100 cm

| Electron Cone (cm×cm) | Field Size (cm×cm) | Output Factor (cGy/MU) | | | | |
|---|---|---|---|---|---|---|
| | | 6 MeV | 9 MeV | 12 MeV | 15 MeV | 18 MeV |
| 6×6 | 6×6 | 1.005 | 1.021 | 0.972 | 1.004 | 1.029 |
| | 5×5 | 1.000 | 1.013 | 0.956 | 0.991 | 1.020 |
| | 4×4 | 1.000 | 0.997 | 0.947 | 0.980 | 1.012 |
| | 3×3 | 0.944 | 0.906 | 0.893 | 0.924 | 0.965 |
| 10×10 | 10×10 | 1.000 | 1.000 | 1.000 | 1.000 | 1.000 |
| | 9×9 | 1.002 | 1.000 | 1.000 | 1.000 | 1.000 |
| | 8×8 | 1.002 | 1.000 | 0.998 | 0.996 | 0.994 |
| | 7×7 | 1.000 | 0.999 | 0.996 | 0.986 | 0.991 |
| | 6×6 | 0.997 | 0.985 | 0.986 | 0.977 | 0.982 |
| | 5×5 | 0.995 | 0980 | 0.981 | 0.972 | 0.969 |
| | 4×4 | 0.950 | 0.913 | 0.942 | 0.940 | 0.960 |
| | 3×3 | 0.903 | 0.831 | 0.902 | 0.899 | 0.922 |
| 15×15 | 15×15 | 0.997 | 0.926 | 1.002 | 0.983 | 0.964 |
| | 14×14 | 1.002 | 0.929 | 1.006 | 0.986 | 0.970 |
| | 12×12 | 1.002 | 0.931 | 1.009 | 0.989 | 0.972 |
| | 10×10 | 1.004 | 0.934 | 1.012 | 0.994 | 0.973 |
| | 7×7 | 1.007 | 0.932 | 1.007 | 0.981 | 0.960 |
| | 5×5 | 0.998 | 0.910 | 0.976 | 0.956 | 0.950 |
| 20×20 | 20×20 | 1.081 | 0.974 | 1.032 | 0.978 | 0.932 |
| | 18×18 | 1.082 | 0.978 | 1.036 | 0.983 | 0.940 |
| | 15×15 | 1.083 | 0.982 | 1.042 | 0.987 | 0.945 |
| | 7×7 | 1.090 | 0.983 | 1.040 | 0.982 | 0.936 |
| 25×25 | 25×25 | 1.057 | 0.967 | 1.023 | 0.953 | 0.903 |
| | 20×20 | 1.058 | 0.972 | 1.027 | 0.969 | 0.920 |
| | 12×12 | 1.064 | 0.978 | 1.038 | 0.969 | 0.925 |

Adapted from Pennington, E.C. and Jani, S.K., The University of Iowa Hospitals and Clinics, Iowa City, Iowa, private communication, 1992.

## TABLE 2.C2
### Effect of Field Shaping on Electron Beam Output for Varian Clinac-2500, SSD=100 cm

| Electron Cone (cm×cm) | Field Size (cm×cm) | Output Factor (cGy/MU) | | | | | |
|---|---|---|---|---|---|---|---|
| | | 6 MeV | 9 MeV | 12 MeV | 15 MeV | 18 MeV | 22 MeV |
| 6×6 | 6×6 | 1.000 | 1.027 | 1.043 | 0.957 | 0.998 | 1.037 |
| | 5×5 | 0.995 | 1.000 | 1.015 | 0.945 | 0.985 | 1.022 |
| | 4×4 | 0.975 | 0.962 | 0.950 | 0.937 | 0.970 | 1.008 |
| | 3×3 | 0.930 | 0.930 | 0.920 | 0.900 | 0.920 | 0.940 |
| 10×10 | 10×10 | 1.000 | 1.000 | 1.000 | 1.000 | 1.000 | 1.000 |
| | 9×9 | 1.000 | 1.000 | 1.000 | 1.000 | 1.000 | 1.000 |
| | 8×8 | 1.000 | 1.000 | 0.996 | 0.996 | 0.996 | 0.996 |
| | 7×7 | 0.980 | 0.995 | 0.990 | 0.990 | 0.990 | 0.987 |
| | 6×6 | 0.965 | 0.980 | 0.955 | 0.976 | 0.978 | 0.978 |
| | 5×5 | 0.950 | 0.970 | 0.940 | 0.970 | 0.970 | 0.963 |
| | 4×4 | 0.950 | 0.960 | 0.930 | 0.940 | 0.960 | 0.950 |
| 15×15 | 15×15 | 0.995 | 0.932 | 0.890 | 0.962 | 0.942 | 0.916 |
| | 10×10 | 1.000 | 0.935 | 0.890 | 0.966 | 0.945 | 0.922 |
| | 7×7 | 1.000 | 0.910 | 0.880 | 0.945 | 0.930 | 0.910 |
| 20×20 | 20×20 | 1.055 | 0.974 | 0.890 | 0.947 | 0.907 | 0.835 |
| | 10×10 | 1.055 | 0.985 | 0.892 | 0.955 | 0.917 | 0.881 |
| 25×25 | 25×25 | 1.027 | 0.991 | 0.883 | 0.931 | 0.880 | 0.835 |
| | 15×15 | 1.032 | 0.991 | 0.883 | 0.947 | 0.896 | 0.855 |

From Pennington, E.C. and Jani, S.K., The University of Iowa Hospitals and Clinics, Iowa City, Iowa, private communication, 1992. With permission.

## TABLE 2.C3
## Effect of Field Shaping on Electron Beam Output for Siemens Mevatron XX, SSD=100 cm

| Electron Cone (cm×cm) | Field Size (cm×cm) | Output Factor (cGy/MU) | | | | | |
|---|---|---|---|---|---|---|---|
| | | 5 MeV | 7 MeV | 10 MeV | 12 MeV | 15 MeV | 18 MeV |
| 10×10 | 10×10 | 1.000 | 1.000 | 1.000 | 1.000 | 1.000 | 1.000 |
| | 5×10 | 0.999 | 0.981 | 1.005 | 1.002 | 1.031 | 1.022 |
| | 5×5 | 0.991 | 0.953 | 0.992 | 0.991 | 1.053 | 1.035 |
| | 4×10 | 1.023 | 1.019 | 1.021 | 1.009 | 1.026 | 1.029 |
| | 4×5 | 1.020 | 0.997 | 1.019 | 1.009 | 1.052 | 1.049 |
| | 4×4 | 1.034 | 1.031 | 1.032 | 1.011 | 1.042 | 1.053 |
| | 3×10 | 1.012 | 0.999 | 0.995 | 1.020 | 1.032 | 1.037 |
| | 3×5 | 1.009 | 0.976 | 0.987 | 1.022 | 1.057 | 1.058 |
| | 3×3 | 1.010 | 0.991 | 0.972 | 1.043 | 1.054 | 1.072 |
| | 2×10 | 0.997 | 0.982 | 0.998 | 1.019 | 1.019 | 1.029 |
| | 2×5 | 0.993 | 0.960 | 0.993 | 1.015 | 1.044 | 1.048 |
| | 2×4 | 1.022 | 0.999 | 1.014 | 1.027 | 1.040 | 1.058 |
| | 2×3 | 1.006 | 0.978 | 0.979 | 1.040 | 1.045 | 1.066 |
| | 2×2 | 0.985 | 0.954 | 0.983 | 1.032 | 1.031 | 1.050 |

From Niroomand-Rad, A., Film dosimetry of small elongated electron beams for treatment planning, *Med. Phys.*, 16, 655, 1989. With permission.

**TABLE 2.C4**

**Effect of Field Shaping on Electron Beam Output for Philips SL-20, SSD=100 cm**

| Electron Cone (cm×cm) | Field Size (cm×cm) | Output Factor (cGy/MU) | | | |
|---|---|---|---|---|---|
| | | 4 MeV | 8 MeV | 12 MeV | 18 MeV |
| 10×10 | 10×10 | 1.000 | 1.000 | 1.000 | 1.000 |
| | 5×5 | 1.018 | 1.012 | 1.012 | 1.031 |
| | 3×3 | 0.977 | 0.935 | 0.991 | 1.039 |
| | 3×9 | 1.002 | 0.970 | 0.999 | 1.032 |
| | 5×9 | 1.015 | 1.008 | 1.010 | 1.033 |
| 20×20 | 20×20 | 1.000 | 1.000 | 1.000 | 1.000 |
| | 16×16 | 1.002 | 1.001 | 1.001 | 1.004 |
| | 12×12 | 1.005 | 1.004 | 1.002 | 1.016 |
| | 8×8 | 1.011 | 1.009 | 1.009 | 1.036 |
| | 4×4 | 1.010 | 0.982 | 1.010 | 1.064 |
| | 3×8 | 1.007 | 0.982 | 1.011 | 1.066 |
| | 5×16 | 1.007 | 1.002 | 1.003 | 1.036 |

From Lamba, M.A.S. and Elson, H.R., The University of Cincinnati Medical Center, Cincinnati, Ohio, private communication, 1992.

## TABLE 2.C5
### Effect of Field Shaping on Electron Beam Output for Philips SL-25, SSD=100 cm

| Electron Cone (cm×cm) | Field Size (cm×cm) | Output Factor (cGy/MU) | | |
|---|---|---|---|---|
| | | 6 MeV | 10 MeV | 20 MeV |
| 10×10 cm | 10×10 | 1.000 | 1.000 | 1.000 |
| | 6×6 | 0.993 | 0.995 | 1.014 |
| | 5×5 | 0.977 | 0.975 | 1.014 |
| | 4×4 | 0.942 | 0.945 | 1.005 |
| | 3×3 | 0.900 | 0.919 | 1.000 |
| | 2×2 | 0.796 | 0.874 | 0.986 |
| | 6×10 | 0.996 | 0.998 | 1.002 |
| | 4×10 | 0.974 | 0.968 | 0.998 |
| | 3×10 | 0.940 | 0.948 | 1.004 |
| | 2×10 | 0.862 | 0.921 | 1.003 |
| | 4×6 | 0.965 | 0.958 | 1.004 |
| | 3×6 | 0.931 | 0.936 | 0.997 |
| | 2×6 | 0.850 | 0.900 | 0.990 |

From Rashid, H., Islam, M.K., Gaballa, H., Rosenow, U.F., and Ting, J.Y., Small-field electron dosimetry for the Philips SL25 linear accelerator, *Med. Phys.*, 17, 710, 1990. With permission.

## TABLE 2.C6
### Effect of Field Shaping on Electron Beam Output for Scanditronix Microtron MM-22, SSD=100 cm

| Field Size (cm×cm) | Output Factor (cGy/MU) | | | | | | | |
|---|---|---|---|---|---|---|---|---|
| | 3 MeV | 5 MeV | 7 MeV | 9 MeV | 13 MeV | 16 MeV | 18 MeV | 22 MeV |
| 25×25 | 0.978 | 1.003 | 0.989 | 0.979 | 0.972 | 0.974 | 0.973 | 0.940 |
| 18×18 | 1.006 | 1.012 | 0.995 | 0.985 | 0.979 | 0.975 | 0.969 | 0.948 |
| 15×15 | 1.019 | 1.012 | 0.994 | 0.987 | 0.985 | 0982 | 0.980 | 0.971 |
| 12×12 | 1.030 | 1.013 | 0.995 | 0.989 | 0.988 | 0.986 | 0.984 | 0.975 |
| 10×10 | 1.000 | 1.000 | 1.000 | 1.000 | 1.000 | 1.000 | 1.000 | 1.000 |
| 8×8 | 1.000 | 0.999 | 0.996 | 0.997 | 0.993 | 0.995 | 0.996 | 0.985 |
| 6×6 | 0.932 | 0.999 | 0.986 | 0.979 | 0.967 | 0.974 | 0.976 | 0.985 |
| 4×4 | 0.900 | 0.928 | 0.913 | 0.893 | 0.888 | 0.918 | 0.936 | 0.965 |
| 9 cm diam. | 0.995 | 1.000 | 0.999 | 0.998 | 0.998 | 0.995 | 0.995 | 0.999 |
| 6 cm diam. | 0.971 | 0.985 | 0.977 | 0.968 | 0.952 | 0.963 | 0.970 | 0.985 |
| 4 cm diam. | 0.874 | 0.891 | 0.880 | 0.878 | 0.857 | 0.894 | 0.918 | 0.961 |
| 2 cm diam. | 0.431 | 0.611 | 0.580 | 0.597 | 0.577 | 0.671 | 0.742 | 0.851 |

From George, R.E., Frost, S.V., and Hartson-Eaton, M., Characteristics of electron beams from a medical microtron, *Med. Phys.*, 13, 533, 1986. With permission.

**TABLE 2.C7**
**Effect of Field Shaping on Electron Beam Output for AECL Therac-20, SSD=100 cm**

| Electron Cone | Field Size | Output Factor (cGy/MU) | | | |
|---|---|---|---|---|---|
| (cm×cm) | (cm×cm) | 6 MeV | 9 MeV | 13 MeV | 17 MeV |
| Scanning beam | 10×10 | 1.000 | 1.000 | 1.000 | 1.000 |
| | 6×6 | 0.886 | 0.948 | 0.955 | 0.980 |
| | 5×5 | 0.843 | 0.922 | 0.936 | 0.966 |
| | 4×4 | 0.762 | 0.859 | 0.871 | 0.960 |

From Mills, M.D., Hogstrom K.R., and Almond, P.R., Prediction of electron beam output factors, *Med. Phys.*, 9, 60, 1982. With permission.

# REFERENCES

1. AAPM, American Association of Physicists in Medicine, *Total Skin Electron Therapy: Technique and Dosimetry*, AAPM Report No. 23 American Institute of Physics, New York, 1988.

2. **Almond, P.R.,** Characteristics of current medical electron accelerator beams, in *Proceedings of the Symposium on Electron Beam Therapy*, Chu, F.C.H. and Laughlin, J. S. Eds., Memorial Sloan Kettering Cancer Center, New York, 1981, 43.

3. **Almond, P.R.,** Radiation Physics of electron beams, in *Clinical Applications of the Electron Beam*, N. Tapley, Ed., Wiley, New York, 1976.

4. **Biggs, P.,** The change in percentage depth dose of electrons due to beam angulation, *Med. Dosimetry*, 9, 25, 1984.

5. **Biggs, P.J., Boyer, A.L., and Doppke, K.P.,** Electron Dosimetry of irregular fields on Clinac-18, *Int. J. Radiat. Oncol. Biol. Phys.*, 5, 433, 1979.

6. **Bova, F.,** A film phantom for routine film dosimetry in the clinical environment, *Med. Dosimetry,* 15, 83, 1990.

7. **Brahme, A., Kraepelien, T., Svensson, H.,** Electron and photon beams from a 50 MeV racetrack microtron, *Acta Radiol. Oncol.*, 19, 305, 1980.

8. **Brahme, A.,** Quality parameters of electron beams from therapy accelerators, Proc. Symp. Electron Dosimetry and Arc Therapy, AAPM American Institute of Physics, 1982.

9. **Brenner, M., Karjalainen, P., Rytila, A., and Jungar, H.,** The effects of inhomogeneities on dose distribution of high-energy electrons, *Ann. N.Y. Acad. Sci.*, 161, 233, 1969.

10. **Bruinvis, I.A.D.,** Dose calculation for arbitrarily shaped electron beams, in *The Computation of Dose Distributions in Electron Beam Radiotherapy*, Nahum, A.E., Ed. Umea University, 1985, 210

11. **Bruinvis, I.A.D., Heukelom, S., and Mijnheer, B.J.,** Comparison of ionization measurements in water and polystyrene for electron beam dosimetry, *Phys. Med. Biol.*, 30, 1043 1985.

12. **Chen, F-S.,** An analytical equation of electron beams percentage depth ionization curve along the central axis, *Med. Phys.*, 15, 407, 1988.

13. **Choi, M.C., Purdy, J.A., Gerbi, B., Abrath, F.G., and Glasgow, G.P.,** Variation in output factor caused by secondary blocking for 7–16 MeV electron beams, *Med. Phys.*, 6, 137, 1979.

14. *Clinical Applications of the Electron Beam*, N. Tapley, Ed., Wiley, New York, 1976.

15. **Ekstrand, K.E., and Dixon, R.L.,** The problem of obliquely incident beams in electron-beam treatment planning, *Med. Phys.*, 9, 276, 1982.

16. **George, R.E., Frost, S.V., and Hartson-Eaton, M.,** Characteristics of electron beams from a medical microtron, *Med. Phys.*, 13, 533, 1986.

17. **Gerbi, B.J., Khan, F.M., Deibel, C., and Kim, T.H.,** Total skin electron arc irradiation using a reclined patient position, *Int. J. Radiat. Oncol. Biol., Phys.*, 17, 397, 1989.

18. **Hettinger, G. and Svensson, H.,** Photographic film for determination of isodoses from betatron radiation, *Acta Radiol.*, 6, 74, 1967.

19. **Hogstrom K.R.**, Dosimetry of electron heterogeneities, in *Advances in Radiation Therapy Treatment Planning (Medical Physics Monograph No. 9)*, Wright, A.E. and Boyer, A.L., Eds., American Institute of Physics, New York, 1983.

20. **Hogstrom, K.R., Mills, M.D., and Almond, P.R.**, Electron beam dose calculations, *Phys. Med. Biol.*, 26, 445, 1981.

21. HPA, Hospital Physicists Association, Code of practice for electron beam dosimetry in radiotherapy, *Phys. Med. Biol.*, 30, 1169, 1985.

22. ICRU, International Commission on Radiation Units and Measurements, Radiation Dosimetry: Electron Beams with Energies Between 1 and 50 MeV, ICRU Report 35 (International Commission on Radiation Units and Measurement, Bethesda, Maryland, 1984.

23. ICRU, International Commission on Radiation Units and Measurements, Radiation Dosimetry: Electrons with Initial Energies Between 1 and 50 MeV, ICRU Report 21 (International Commission on Radiation Units and Measurement, Bethesda, Maryland, 1972.

24. **Jamshidi, A., Kuchnir F.T., and Reft, C.S.**, Determination of the source position for the electron beams from a high-energy linear accelerator, *Med. Phys.*, 13, 942, 1986.

25. **Jamshidi, A., Kuchnir, F.T., and Reft, C.S.**, Characteristic parameters of 6–22 MeV electron beams from a 25 MeV linear accelerator, *Med. Phys.*, 14, 282, 1987.

26. **Karlsson, M., Nystrom H., and Svensson, H.**, Electron beam characteristics of the 50-MeV racetrack microtron, *Med. Phys.*, 19, 307, 1992.

27. **Khan, F.M., Deibel, F.C., and Soleimani-Meigooni, A.**, Obliquity correction for electron beams, *Med. Phys.*, 12, 749, 1985.

28. **Khan, F.M., Doppke, K.P., Hogstrom, K.R., Kutcher, G.J., Nath, R., Prasad, S.C., Purdy, J.A., Rozenfeld, M., and Werner, B.L.**, Clinical electron-beam dosimetry: Report of AAPM Radiation Therapy Committee Task Group No. 25, *Med. Phys.*, 18, 73, 1991.

29. **Khan, F.M., Moore, V.C., and Levitt, S.H.**, Field shaping in electron beam therapy, *Br. J. Radiol.*, 49, 883, 1976.

30. **Khan, F.M., Sewchand, W., and Levitt, S.H.**, Effect of air space on depth dose in electron beam therapy, *Radiology*, 126, 249, 1978.

31. **Khan, F.M., Werner, B.L., and Deibel, F.C.**, Lead shielding for electrons, *Med. Phys.*, 8, 712, 1981.

32. **Kirby, T.H., Gastorf, R.J., Hanson, W.F., Berkley, L.W., Gagnon, W.F., Hazle, J.D., and Shalek, R.J.**, Electron beam central axis depth dose measurements, *Med. Phys.*, 12, 357, 1985.

33. **Klevenhagen, S.C.**, Physics of Electron Beam Therapy, Adam Hilger, Ltd., Bristol and Boston, 1985.

34. **Lamba, M.A.S. and Elson, H.R.**, The University of Cincinnati Medical Center, Cincinnati, Ohio, private communication, 1992.

35. **Lax, I. and Brahme, A.**, Collimation of high energy electron beams, *Acta Radiol. Oncol.* 19, 199, 1980.

36. **Leavitt, D., Peacock, L., Gibbs, Jr., F., and Stewart, J.**, Electron arc therapy: physical measurements and treatment planning techniques, *Int. J. Radiat. Oncol. Biol. Phys.*, 11, 987, 1985.

37. **Leavitt, D.D., Stewart, J.R., and Early, L.,** Improved dose homogeneity in electron arc therapy achieved by a multiple energy technique, *Int. J. Radiat. Oncol. Biol. Phys.,* 19, 159, 1990.

38. **McGinley, P.H., McLaren, J.R., and Barnett, B.R.,** Small electron beams in radiation therapy, *Radiology,* 131, 231, 1979.

39. **McParland, B.J.,** A method of calculating the output factors of arbitrarily shaped electron fields, *Med. Phys.,* 16, 88, 1989.

40. **McParland, B.J.,** A parameterization of the electron beam output factors of a 25-MeV linear accelerator, *Med. Phys.,* 14, 665, 1987.

41. **Meigooni, A.S. and Das, I.J.,** Parametrization of depth dose for electron beams, *Phys. Med. Biol.,* 32, 761, 1987.

42. **Meyer, J.A., Palta, J.R., and Hogstrom K.R.,** Demonstration of relatively new electron dosimetry measurement techniques on the Mevatron 80, *Med. Phys.,* 11, 670, 1984.

43. **Mills, M.D., Hogstrom K.R., and Almond, P.R.,** Prediction of electron beam output factors, *Med. Phys.,* 9, 60, 1982.

44. **Mills, M.D., Hogstrom, K.R., and Fields, R.S.,** Determination of electron beam output factors for a 20 MeV linear accelerator, *Med Phys.,* 12, 473, 1985.

45. NACP, Recommendations by the Nordic Association of Clinical Physics, Procedures in external radiation therapy dosimetry with electron and photon beams with maximum energies between 1 and 50 MeV, *Acta Radiol. Onc.,* 19, 55, 1980.

46. NACP, Supplement to the recommendations by the Nordic Association of Clinical Physics, Electron beams with mean energies at the phantom surface below 15 MeV, *Acta Radiol. Oncol. Rad. Ther. Phys. Biol..,* 20, 402, 1981.

47. **Nair, R.P., Nair, T.K.M., and Wrede, D.E.,** Shaped field electron dosimetry for a Philips SL75/10 linear accelerator, *Med. Phys.,* 10, 356, 1983.

48. **Niroomand-Rad, A.,** Film dosimetry of small elongated electron beams for treatment planning, *Med. Phys.,* 16, 655, 1989.

49. **Niroomand-Rad, A., Gillin, M.T., Kline, R.W., and Grimm, D.F.,** Film dosimetry of small electron beams for routine radiotherapy planning, *Med. Phys.,* 13, 416, 1986.

50. **O'Brien, P., Michaels, H.B., Aldrich J.E., and Andrew, J.W.,** Characteristics of electron beams from a new 25-MeV linear accelerator, *Med. Phys.,* 12, 799, 1985.

51. **Okumura, Y.,** Correction of dose distribution for air space in high-energy electron beam therapy, *Radiology,* 103, 183, 1972.

52. **Palta, J.R., Daftari, I.K., Ayyanger, K.M., and Suntharalingam, N.,** Electron beam characteristics on a Philips SL25, *Med. Phys.,* 17, 27, 1990.

53. **Pennington, E.C. and Jani, S.K.,** The University of Iowa Hospitals and Clinics, Iowa City, Iowa, private communication, 1992.

54. **Pfalzner, P.M. and Clarke, H.C.,** Radiation parameters of 6 to 20 MeV scanning electron beams from the Saturne linear accelerator, *Med. Phys.,* 9, 117, 1982.

55. **Pla, M., Pla, C., and Podgorsak, E.,** The influence of beam parameters on percentage depth dose in electron arc therapy, *Med. Phys.,* 15, 49, 1988.

56. *Practical Aspects of Electron Beam Treatment Planning*, Orton, C.G. and Bagne, F., Eds., Medical Physics Monograph No. 2, American Institute of Physics, N.Y., 1978.

57. **Prasad, S., Ames, T.E., Howard, T.B., Bassano, D.A., Chung, C.T., King, G.A., and Sagerman, R.H.**, Dose enhancement in bone in electron beam therapy, *Radiology*, 151, 513, 1984.

58. **Prasad, S.C., Bedwinek, J.M., and Gerber, R.L.**, Lung dose in electron beam therapy of chest wall, *Acta Radiol. Oncol.*, 22, 91, 1983.

59. Proceedings of the Symposium on Electron Dosimetry and Arc Therapy, Paliwal, B., Ed., American Institutes of Physics, Inc., New York, 1982.

60. **Purdy, J.A., Abrath, F.G., and Bello, J.E.**, Electron dosimetry for shaped fields on the Clinac-20, In *Proceedings of the Symposium on Electron Dosimetry and Arc Therapy*, Paliwal, B., Ed., American Institute of Physics, New York, 1982, 327.

61. **Purdy, J.A., Choi, M.C., and Feldman, A.**, Lipowitz metal shielding thickness for dose reduction of 6–20 MeV electrons, *Med. Phys.*, 7, 251, 1980.

62. **Rashid, H., Islam, M.K., Gaballa, H., Rosenow, U.F., and Ting, J.Y.**, Small-field electron dosimetry for the Philips SL25 linear accelerator, *Med. Phys.*, 17, 710, 1990.

63. **Rogers, D.W.O. and Bielajew, A.F.**, Differences in electron depth dose curves calculated with EGS and ETRAN and improved energy-range relationships, *Med. Phys.*, 13, 687, 1986.

64. **Ruegsegger, D., Lerude, S., and Dick, L.**, Electron beam arc therapy using a high energy betatron, *Radiology*, 133, 483, 1979.

65. **Rustgi, S.N. and Rodgers, J.E.**, Analysis of bremsstrahlung component in 6–18 MeV electron beams, *Med. Phys.*, 14, 884, 1987.

66. **Schroder-Babo, P.**, Determination of the virtual electron source of a betatron, *Acta Radiol. Suppl.*, 364, 7, 1983.

67. **Sharma, S.C. and Wilson, D.L.**, Depth dose characteristics of elongated fields for electron beams from a 20 MeV accelerator, *Med. Phys.*, 12, 419, 1985.

68. **Sharma, S.C., Wilson, D.L., and Jose, B.**, Dosimetry of small fields for therac 20 electron beams, *Med. Phys.*, 11, 697, 1984.

69. **Shiu, A.S., Otte, V.A., and Hogstrom K.R.**, Measurement of dose distributions using film in therapeutic electron beams, *Med. Phys.*, 16, 911, 1989.

70. **Shortt, K.R., Ross, C.K., Bielajew, A.F., and Rogers, D.W.O.**, Electron beam dose distributions near standard inhomogeneities, *Phys. Med. Biol.*, 31, 235, 1986.

71. **Spira, J., Botstein, C., Eisenberg, B., and Berdon, B.**, Betatron: electron beam 10–35 MeV. Central depth doses and isodoses curves, *American J. Roentgen.*, 88, 262, 1962.

72. **Ten Haken, R.K. and Fraas, B.A.**, Determination of electron beam mean incident energy from $D_{50}$ (ionization) values, *Med. Phys.*, 14, 985, 1987.

73. **Ten Haken, R.K. and Fraass, B.A.**, Relative electron beam measurements: scaling depths in clear polystyrene to equivalent depths in water, *Med. Phys.*, 14, 410, 1987.

74. **Ten Haken, R.K., Fraass, B.A., and Jost, R.J.,** Practical methods of electron depth-dose measurement compared to use of the NACP design chamber in water, *Med. Phys.,* 14, 1060, 1987.

75. **Thomadsen, B.R., Asp, L.W., Van de Geijn, J., Paliwal, B.R. and Po-Cheng, C.,** Perturbation of electron beam doses as a function of SSD due to the use of shielding blocks on the Clinac-18, *Med. Phys.,* 8, 507, 1981.

76. **Thomas, S.J.,** Virtual source distances for electron beams between 5 and 20 MeV, *Phys. Med. Biol.,* 33, 1325, 1988.

77. **Udale, M.A.,** A Monte Carlo investigation of surface doses for broad electron beams, *Phys. Med. Biol.,* 33, 939, 1988.

78. **Van der Laarse, R., Bruinvis, I.A.D., and Nooman, M.F.,** Wall-scattering effects in electron beam collimation, *Acta Radiol. Oncol.,* 17, 113, 1978.

79. **VanBattum, L.J. and Huizenga, H.,** Film dosimetry of clinical electron beams, *Int. J. Radiol. Oncol. Biol. Phys.,* 18, 69, 1990.

80. **Wu, A., Kalend, A.M., Zicker, R.D., and Sternick, E.S.,** Comments on the method of energy determination for electron beams in the TG-21 protocol, *Med. Phys.,* 11, 871, 1984.

Chapter 3

# DATA ON SPECIAL RADIOTHERAPEUTIC PROCEDURES

Certain radiotherapeutic procedures require special attention in developing the treatment techniques, performing dosimetric measurements, and calibrating the radiation beams. Often, the treatment units are modified to achieve unique beam geometry for specific radiotherapeutic goals. In such circumstances, any previously published reports on the subject matter become quite useful. In this chapter, we present a limited amount of dosimetric data for four special procedures:

A. Total Body Photon Irradiation (TBI);
B. Total Skin Electron Therapy (TSET);
C. Stereotactic Radiosurgery (SRS); and
D. Intraoperative Radiotherapy (IORT).

For each of the above techniques, we have presented some sample data tables, and figures. At the end of this chapter, a detailed bibliography is provided separately for each of the above four topics.

## A. TOTAL BODY PHOTON IRRADIATION (TBI)

The TBI procedure requires a radiation field large enough to intercept an adult patient's entire body. An easy way to obtain such a field is to treat the patient at an extended distance (often 3–5 times the standard source-to-isocenter distance of 0.8–1.0 meter) with horizontal beam geometry. The machine collimators can be rotated 45° from normal settings to utilize the diagonal field length. Usually, the TBI procedure is part of an overall treatment regimen which includes chemotherapy. And so, the timing of the TBI treatment becomes very critical. To insure this, one wisely chooses a very simple, reliable treatment machine, such as a Co-60 teletherapy unit. Numerous papers have appeared in the literature on TBI techniques using cobalt units, either in standard or modified configuration. Some attempts have also been made to manufacture the treatment units dedicated solely to TBI procedures.[23] The University of Iowa Co-60 machine is modified with a special collimating assembly and a beam flattening filter to achieve a vertical beam geometry.[18] The central axis percent depth dose data for TBI fields from Co-60 machines are listed in Table 3.A1. Data show that for parallel-opposed, lateral field treatments in horizontal beam geometry, the cobalt beam penetration is inadequate to assure a uniform dose to the entire body. For bilateral treatments, it is advantageous to employ high energy photon beams. The central axis depth dose data for 4–25-MV X ray beams are listed in Table 3.A2.

During TBI procedures, it is vital to achieve full dose to the skin. Large field geometry inherently causes a rapid buildup of dose with depth, but often not enough to remove entire skin-sparing effect of these megavoltage beams. Therefore, a beam-spoiler plate or screen is commonly placed close to the patient during TBI treatments to achieve full electronic equilibrium at or near the skin. Table 3.A3 lists typical surface dose values for various beam energies with TBI techniques.

Large radiation fields from cobalt beams usually exhibit forward-peaking intensity. In other words, the intensity is usually maximum at the central ray and falls off as one moves away towards the field edge. A beam flattening filter is normally placed within the beam to ensure its flatness and, hence, a very uniform dose throughout the body with parallel-opposed treatments. For linear accelerator beams, the intensity at off-axis points can be higher than the central axis due to radiation horns. The user must take into account such variations in beam intensity throughout the large TBI fields. Figures 3.1 through 3.4 show typical beam profiles for TBI fields from various treatment units. The profile labelled "open field" in Figure 3.1 represents cobalt machine geometry where its own standard collimating device was replaced by a special collimating block which resulted in a $260 \times 75$ cm$^2$ field size at about 165 cm SSD. Figures 3.3 and 3.4 show that dose profiles across TBI fields do not change appreciably with distance when geometric beam divergence is considered.

An excellent reference on this topic is the AAPM Report-17 that describes in detail the dosimetric aspects of a successful TBI procedure.[1] We believe that the user:

1.  must measure dosimetry data at extended SSD *under* TBI conditions;
2.  must measure doses throughout the field to assure dose uniformity;
3.  must use "full" phantoms as inadequate phantom size could affect dosimetry data;[32]
4.  must verify doses in a full (preferably human body) phantom using detectors other than those used for calibrations; and, finally,
5.  should refer to published reports for data comparison to avoid blunders!

## B.  TOTAL SKIN ELECTRON THERAPY (TSET)

The TSET is a very complex treatment technique in which the entire skin (cutaneous tissues) of the patient is treated to a more or less uniform dose using multiple electron beam arrangement. The beam energy used is about 5–10 MeV in range at the accelerator window. Since very large treatment fields are needed, the patient is usually placed at an extended distance (2–4 meter) from the source. A lucite (or Plexiglas) screen is placed in front of the patient intercepting the electron beam before it reaches the patient. This assures that the skin dose is near maximum and that no skin-sparing (i.e., buildup of dose within tissues) occurs. The effective beam energy at the patient may be reduced to 3–7 MeV. Figure 3.5 shows the effect of plastic beam spoiler on the depth dose of a broad electron beam.

The intensity of electron beams with large field size and extended SSD may be quite non-uniform. Therefore, when treating the entire skin surface, several fields may have to be used with some sort of field edge matching arrangement. Figure 3.6 shows the dose profiles in phantom for a large electron field as well as for two matched electron fields. If the treatment is to be given with a parallel-opposed anterior-posterior field arrangement, a portion of the lateral skin receives too little a dose. This is illustrated in Figure 3.7(a). As the number of fields directed at the patient from around increases, the penetration (depth dose) of the beam becomes uniform all around the patient. This is illustrated in Figures 3.7(b), (c), and (d).

The original Stanford technique for TSET employed four body orientations, anterior, posterior, and two lateral fields of a standing patient.[16] To obtain a better dose uniformity, this technique was later modified,[73] and the setup was replaced by a six-field technique. It involved an anterior, posterior and four oblique fields. The depth-dose (beam penetration) resulting from a six-field TSET decreases compared to a single field depth-dose. This is true for most multi-field extended SSD techniques or electron arc therapy.

AAPM Report-23 is an excellent reference on the TSET technique and dosimetry.[45] It provides details on basic requirements for TSET therapy, methods of dosimetry and various treatment techniques.

## C.  STEREOTACTIC RADIOSURGERY (SRS)

Stereotactic radiosurgery is an irradiation technique that precisely locates and treats intracranial lesions to a very high dose. Originally, this technique, pioneered by Leksell, utilized a treatment unit called Gamma Knife.[109] A Gamma Knife housed 201 Co-60 sources arranged in a hemispherical dome. All the beams from these sources were individually collimated to focus at a point within the dose. With a stereotactic frame attached to the patient's head, the target lesion was precisely located. The stereotactic frame also assisted in positioning the lesion within the dome of a Gamma Knife to a point where all the cobalt beams converge. The Gamma Knife offered a precision of about 0.1 mm during the treatment.

Linear accelerator-based stereotactic radiosurgery has been developed in recent years, primarily to avoid huge costs involved with Gamma Knife facilities.[97, 112, 126, 132] Various approaches have been taken to obtain small, very well-defined radiation beams from a medical linear accelerator (linac). Since most all the linacs now are isocentric (rotational) units, the radiation beam can be made to rotate around the patient's head in specific predetermined arcs, thus, simulating a Gamma Knife treatment. Numerous reports have appeared on this subject in the literature in the last decade. The reader is referred to the bibliography given at the end of this chapter for a description of the various techniques and innovative approaches employed in treating brain lesions with a SRS procedure.

The small field dosimetry is complex and requires a great deal of care in carrying out measurement. Rice et al. have evaluated methods to perform small field dosimetry.[129] Commonly used beams range from 6–10 MV in energy. Tables 3.C1 through 3.C8 contain the tissue phantom ratios for small circular fields in the energy range of 4–24 MV X rays. After carefully collecting dosimetry data in phantoms, the user must verify the data by performing tests in a human body phantom using dosimeters other than those employed in original calibration.

## D. INTRAOPERATIVE RADIOTHERAPY (IORT)

Linac based intraoperative radiotherapy is a relatively new treatment technique in which a high single dose of radiation (10–20 Gy) is delivered to the tumor during its direct access at the time of surgery. Electron beams in the energy range of 4–20 MeV are well collimated so as to treat sharply defined tumor volumes. Various types of electron applications (cones) have been developed to achieve this highly localized delivery of radiation. The electron beams of 4–20 MeV energy exhibit two important properties that make them useful for IORT: (1) excellent homogeneity of radiation dose deposited in a selected depth of tissues; and, (2) sharp dose fall-off beyond this depth that spares the underlying normal tissues from unnecessary irradiation.

The dosimetric properties of IORT electron beams depend significantly on the application design. Therefore, a full dosimetric evaluation of these beams is warranted prior to their implementation. The following parameters are usually obtained for each IORT cone size at all available electron energies:

1. Central axis depth dose data
   A. Beam flatness and symmetry;
   B. Beam width;
   C. Beam penumbra;
   D. Isodose curves (needed especially for beveled cones).

2. Beam output (Gy/MU)
3. Variation of output with distance (SSD)
4. Cross-beam data
   A. Beam flatness and symmetry;
   B. Beam width;
   C. Beam penumbra;
   D. Isodose curves (needed especially for beveled cones).

These dosimetric data, especially the beam output values, should be verified with an independent dosimetry detector, such as thermoluminescent dosimeters (TLD).

Typical central axis electron depth dose data for flat-ended IORT cones are given in Table 3.D1. The depth dose data for 15° and 30° beveled IORT cones are given in Tables 3.D2 and 3.D3, respectively. The applicator system for these data sets is very similar to the one described by McCullough and Anderson.[157] The system includes a portion of the applicator common to all cone sizes that is mounted on treatment unit's accessory slot. This common "mother" cone then accepts cylindrical lucite inserts of different diameters (4–9 cm). The resulting source-to-end of the cone distance is 118 cm. The data are valid only for an X ray collimator setting of 10×10 cm². The beam flatness changes quite rapidly with X ray collimator setting—a phenomenon well-documented in the literature. The depth dose data in Tables 3.D1 through 3.D3 represent average values of all cone sizes ranging from 5–9 cm in diameter. The smallest cone size of 4 cm exhibit values significantly lower than the rest of the cones. The surface dose ($D_s$) and X ray contamination dose ($D_x$) are also listed along with the depth doses for IORT cones. Full definitions on $D_s$ and $D_x$ are given in Chapter 2. The electron beam outputs for these three sets of cones are given in Tables 3.D4, 3.D5, and 3.D6.

Beam uniformity across an IORT field throughout the treatment area is affected by the X ray collimator settings.[157] Figure 3.8 shows isodose distributions from a typical cone size for a flat-end, a 15° bevel, and a 30° bevel. The beam penetration alters significantly with the bevel.

## LIST OF FIGURES

## LIST OF TABLES

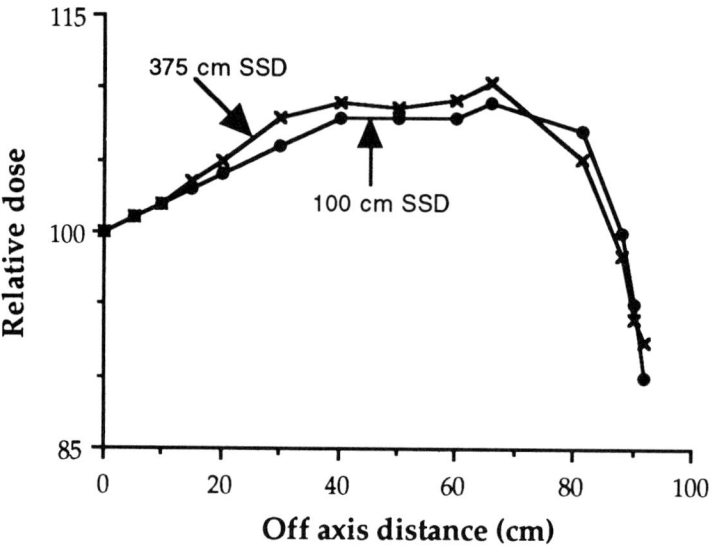

**FIGURE 3.1**: Dose profiles in full phantom across a TBI field from an AECL T-80 cobalt unit. Vertical beam geometry, SSD=167 cm, depth=0.5 cm. See text for details (Ref. 16, 18).

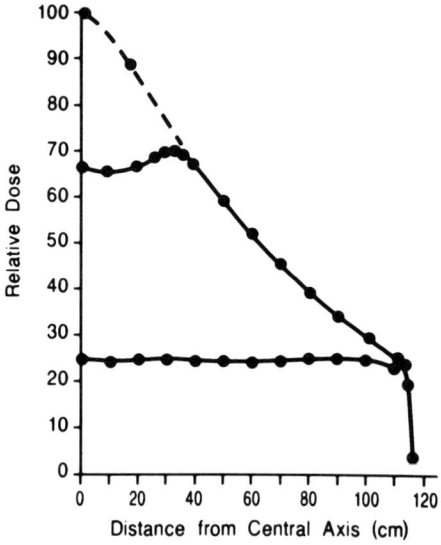

**FIGURE 3.2**: Beam profiles measured in air without any beam filter (upper curve), with the brass-lead filter (middle curve) and with, in addition, the copper flattening filter (lower curve). Varian Clinac-4, 4-MV X rays, 205-cm SSD. (Reprinted from Lutz, W.R., Dougan, P.W., and Bjarngard, B.E., Design and characteristics of a facility for total body and large field irradiation, *Int. J. Radiat. Oncol. Biol. Phys.*, 15, 1035, 1988. With permission.)

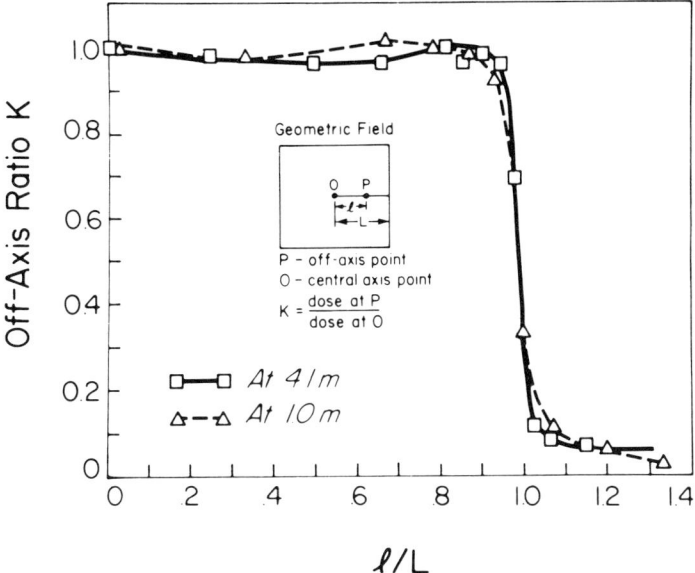

**FIGURE 3.3**: Off-axis ratio ($k$) vs. normalized lateral distance ($l/L$) at 1 m and 4.1 m SSD. Toshiba LMR-13, 10-MV X rays, depth=2.5 cm. (Reprinted from Khan, F.M., Williamson, J.F., Sewchand W., and Kim, T.H., Basic data for dosage calculation and compensation, *Int. J. Radiat. Oncol. Biol. Phys.*, 6, 745, 1980. With permission.)

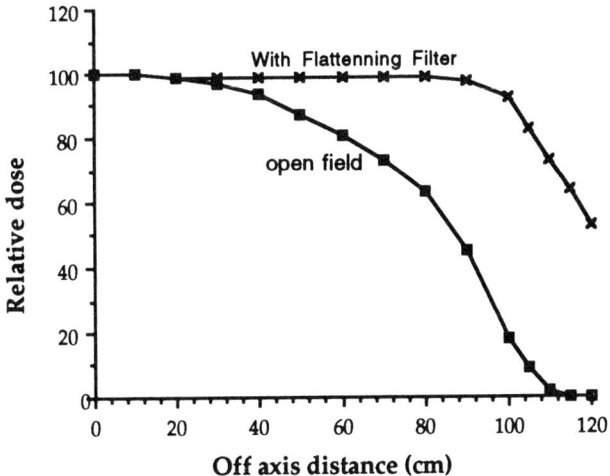

**FIGURE 3.4**: Dose profiles in full phantom across a diagonal direction of a 24-MV TBI field. Varian Clinac-2500. Depth of measurement=5 cm. The 100 cm SSD data are plotted by projecting the off-axis distance at 375 cm (Ref. 16, 17).

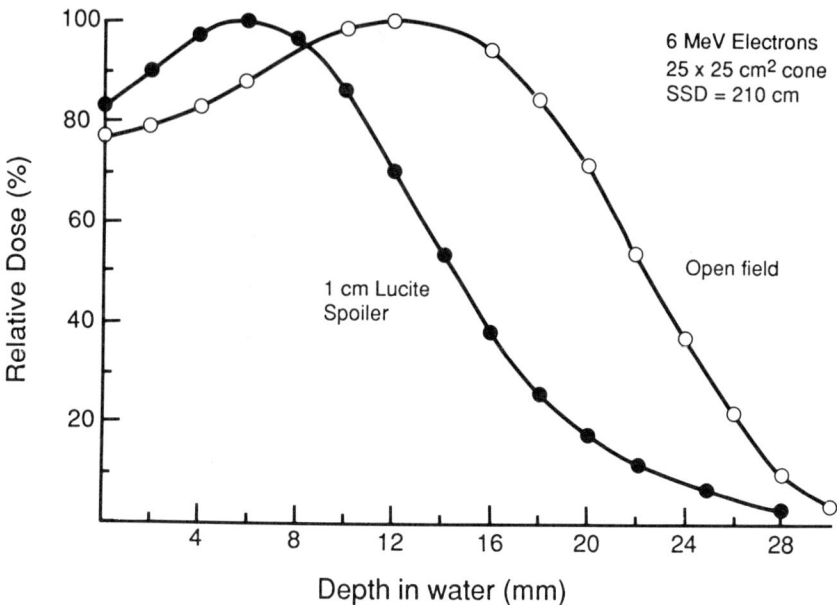

**FIGURE 3.5**:   Depth dose of a 6-MeV electron beam at 210-cm SSD, Varian Clinac-18. The plastic degrader significantly alters the depth dose.

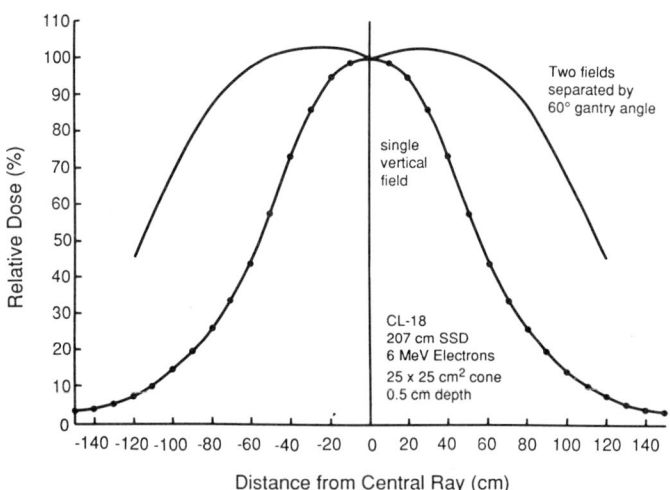

**FIGURE 3.6:**   Dose profiles in phantom at a depth of 0.5 cm resulting from a single large electron field and a pair of fields separated by about 60 cm. Beams are degraded with 1-cm lucite plate.

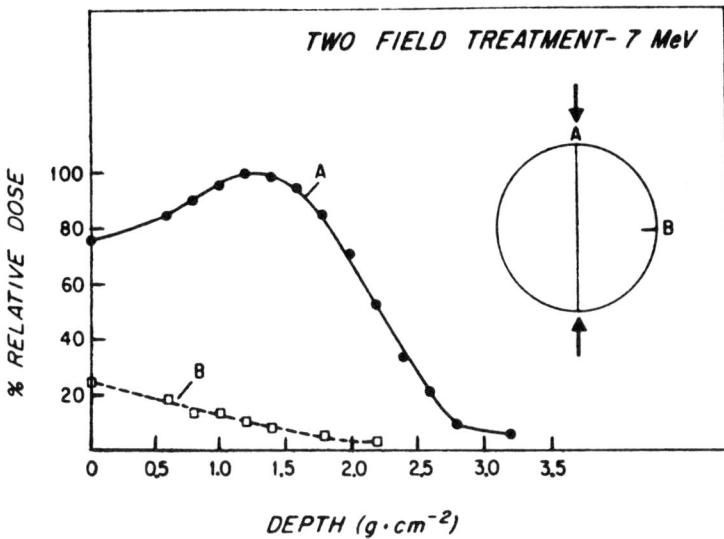

**FIGURE 3.7(a):** Composite depth dose curves for a two field electron treatment, normalized at maximum dose along A. Siemens Mevatron XII, 300-cm SSD. (Reprinted from Bjarngard, B.E., Chen, G.T.Y., Piontek, R.W., and Svensson, G.K., Analysis of dose distributions in whole body superficial electron therapy, *Int. J. Radiat. Oncol. Biol. Phys.*, 2, 319, 1977. With permission.)

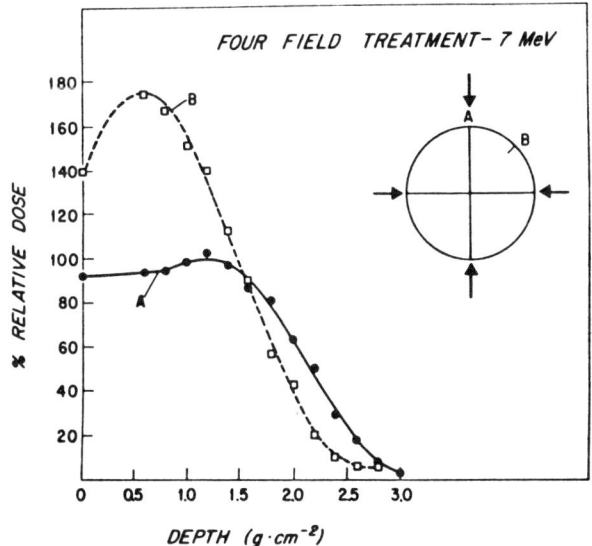

**FIGURE 3.7(b):** Composite depth dose curves for a four-field treatment.

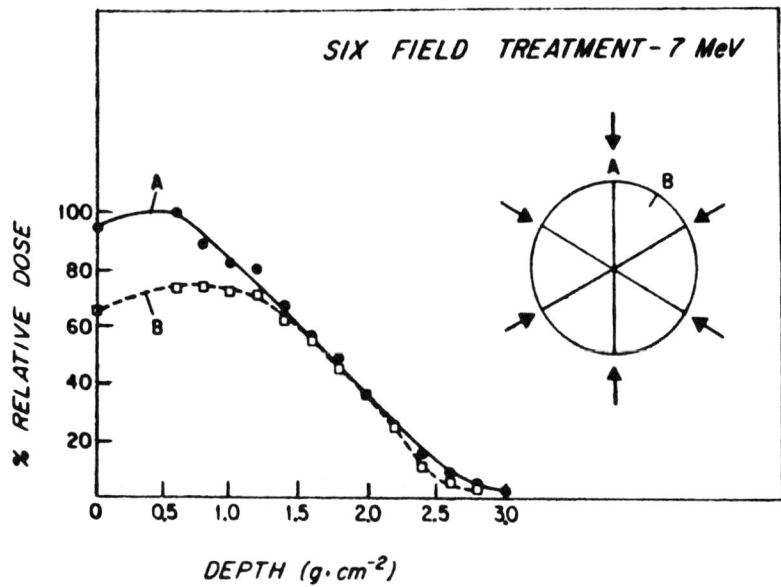

**FIGURE 3.7(c):**    Composite depth dose curves for a six-field treatment.

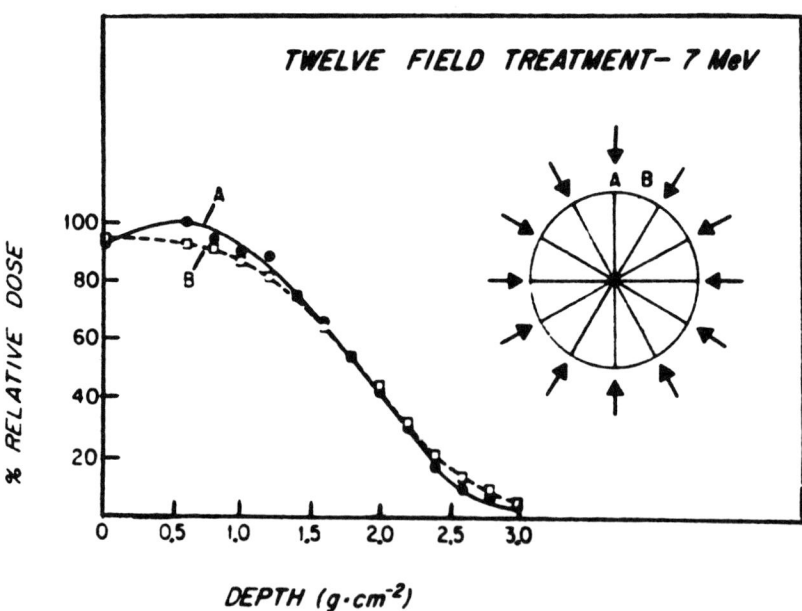

**FIGURE 3.7(d):**    Composite depth dose curves for a twelve-field treatment.

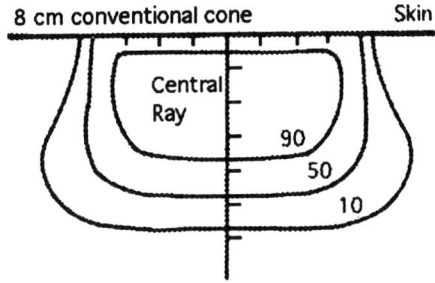

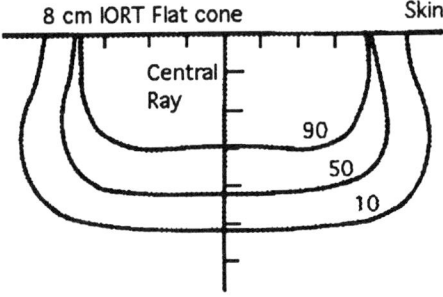

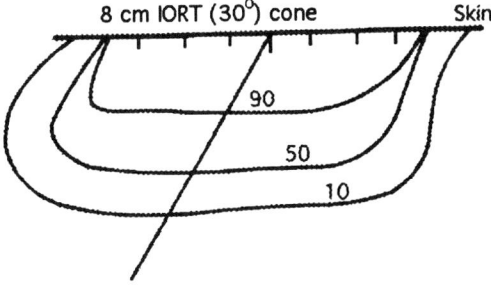

**FIGURE 3.8:** Isodose distribution in water for an IORT flat and 30° beveled cone along with the isodoses of a conventional cone. 118-cm SSD, 12-MeV electron beam from a Varian Clinac-2500.

**TABLE 3.A1**

**Central Axis Depth Dose Data in Full Phantom for TBI Fields from Cobalt Units**

| Depth (cm) | Cobalt-60 Percent Depth Dose | | | |
| --- | --- | --- | --- | --- |
| | SSD=90 cm | SSD=165 cm | SSD=330 cm | SSD=350 cm |
| 0.5 | 100.0 | 100.0 | 100.0 | 100.0 |
| 5.0 | 85.5 | 87.4 | 89.3 | 90.2 |
| 10.0 | 66.8 | 70.5 | 73.3 | 76.0 |
| 15.0 | 51.2 | 55.1 | 58.2 | — |
| 16.5 | — | 51.3 | 54.3 | 57.9 |
| 20.0 | 38.7 | 42.9 | 45.1 | 49.8 |
| **Remarks:** | | | | |
| Machine | Special unit | Modified AECL T-80 | AECL T-80 | AECL T-780 |
| Tx-geometry | Vertical beam | Vertical beam | Horizontal beam | Horizontal beam |
| Field size | 160×50 cm$^2$ | 260×75 cm$^2$ | 135×135 cm$^2$ | Maximum coll. setting |
| Beam modifier | Cu-electron filter | Cu-flattening filter | 1.25-cm lucite spoiler | Bolus at skin |
| Ref. | 42 | 18 | 16 | 32 |

**TABLE 3.A2**
**Central Axis Depth Dose Data in Full Phantom for TBI Fields at Various X Ray Energies**

| Depth (cm) | Percent Depth Dose | | | | | | | |
|---|---|---|---|---|---|---|---|---|
| | 4 MV 190-cm SSD | 4 MV 195-cm SSD | 6 MV 400-cm SSD | 10 MV 300-cm SSD | 10 MV 370-cm SSD | 18 MV 350-cm SSD | 24 MV 300-cm SSD | 24 MV 375-cm SSD |
| 1.0 | 100.0 | 100.0 | — | — | — | 98.0 | 70.0 | 103.6 |
| 1.5 | — | — | 100.0 | — | — | 99.0 | 81.5 | 104.5 |
| 2.5 | — | — | — | 100.0 | 100.0 | 100.0 | 95.0 | 104.3 |
| 3.0 | — | — | — | — | — | 100.0 | 98.0 | 103.5 |
| 5.0 | 90.0 | 90.1 | 93.0 | 96.3 | 97.0 | 97.0 | 100.0 | 100.0 |
| 10.0 | 76.0 | 75.2 | 80.8 | 84.6 | 86.5 | 87.0 | 91.0 | 90.0 |
| 15.0 | 63.0 | 62.0 | 69.0 | 72.8 | 76.0 | 77.0 | 81.7 | 79.8 |
| 20.0 | 51.0 | 48.3 | 58.3 | 63.1 | 66.0 | 67.0 | 73.0 | 69.0 |
| **Remarks:** | | | | | | | | |
| Accelerator | Therapi 4 | Clinac-4 | EMI-6 | Clinac-18 | Clinac-18 | Mevatron XX | Clinac-2500 | Clinac-2500 |
| Tx-geometry | Sweeping vertical beam | Special unit vertical beam | 90° coll. tilt horizontal beam | 90° coll. tilt horizontal beam | 90° coll. tilt horizontal beam | 90° coll. tilt horizontal beam | 90° coll. tilt horizontal beam | 90° coll. tilt horizontal beam |
| Field size | 71×71 cm² | 40×115 cm² | 50×140 cm² | 50×140 cm² | 130×130 cm² | 140×140 cm² | 120×120 cm² | Max. coll. setting |
| Beam modifier | No spoiler | No spoiler | Bolus | Bolus | 0.8-cm bolus | 1-cm perpex | No spoiler | 1.2-cm lucite |
| Ref. | 31 | 24 | 32 | 32 | 8 | 28 | 17 | 16 |

**TABLE 3.A3**
**Surface (Skin) Dose for TBI Fields**

| X Ray Energy | SSD (cm) | Surface Dose (%) for Single TBI Field without Spoiler | Beam Spoiler Employed | Skin Dose (% of TBI dose) from All Fields | Reference |
|---|---|---|---|---|---|
| Co-60 | 165 | 67 | Two blankets | 90 | 18 |
|  | 330 | 87 |  | 90 | 16 |
|  | 350 | 80 | Bolus | ~100 | 32 |
| 4 MV | 195 | 63 | 1.6-mm fabric | 88 | 24 |
| 6 MV | 400 | 65 | Bolus | ~100 | 32 |
| 10 MV | 300 | 40 | Bolus | ~100 | 32 |
|  | 370 | 57 | 16 blankets | 90 | 8 |
|  | 410 | 44 | 0.95-cm lucite | 90 | 35 |
| 18 MV | 350 | 56 | 1-cm perpex | >90 | 28 |
| 24 MV | 375 | 46 | 1.2-cm lucite | 99 | 16 |

## TABLE 3.C1
### Tissue Phantom Ratios for 4-MV Stereotactic Beams [Varian Clinac 4/80, 80-cm SAD]

| Depth (cm) | Field Diameter (cm) | | | | |
|---|---|---|---|---|---|
| | 1.0 | 1.5 | 2.0 | 2.5 | 3.0 |
| 0.5 | 0.961 | 0.949 | 0.945 | 0.943 | 0.942 |
| 1.0 | 0.999 | 1.000 | 1.000 | 1.000 | 1.000 |
| 2.0 | 0.951 | 0.960 | 0.965 | 0.968 | 0.971 |
| 3.0 | 0.901 | 0.911 | 0.918 | 0.924 | 0.929 |
| 4.0 | 0.850 | 0.861 | 0.869 | 0.877 | 0.884 |
| 5.0 | 0.802 | 0.814 | 0.824 | 0.833 | 0.842 |
| 6.0 | 0.756 | 0.768 | 0.779 | 0.788 | 0.798 |
| 7.0 | 0.712 | 0.725 | 0.736 | 0.746 | 0.755 |
| 8.0 | 0.671 | 0.685 | 0.696 | 0.706 | 0.716 |
| 9.0 | 0.632 | 0.646 | 0.657 | 0.668 | 0.678 |
| 10.0 | 0.596 | 0.610 | 0.621 | 0.632 | 0.641 |
| 12.0 | 0.531 | 0.544 | 0.555 | 0.565 | 0.574 |
| 14.0 | 0.473 | 0.485 | 0.496 | 0.506 | 0.515 |
| 16.0 | 0.422 | 0.433 | 0.443 | 0.453 | 0.461 |
| 18.0 | 0.376 | 0.388 | 0.397 | 0.405 | 0.413 |
| 20.0 | 0.336 | 0.347 | 0.356 | 0.363 | 0.370 |
| cGy/MU* | 0.850 | 0.890 | 0.900 | 0.910 | 0.920 |

* Output at 1.0-cm depth; normalized to $10 \times 10$ cm$^2$.

From Serago, C.F., Houdek, P.V., Hartmann, G.H., Saini, D.S., Serago, M.E., and Kaydee, A., Tissue maximum ratios (and other parameters) of small circular 4, 6, 10, 15 and 24 MV x-ray beams for radiosurgery, *Phys. Med. Biol.*, 37, 1943, 1992. With permission.

## TABLE 3.C2
### Tissue Phantom Ratios for 6-MV Stereotactic Beams [Varian Clinac 6/100, 100-cm SAD]

| Depth (cm) | Field Diameter (cm) | | | | | | |
|---|---|---|---|---|---|---|---|
| | 1.25 | 1.5 | 2.0 | 2.5 | 3.0 | 3.5 | 4.0 |
| 0.5 | 0.921 | 0.905 | 0.889 | 0.883 | 0.881 | 0.881 | 0.881 |
| 1.0 | 1.000 | 0.999 | 0.994 | 0.991 | 0.989 | 0.989 | 0.988 |
| 1.5 | 1.000 | 1.000 | 1.000 | 1.000 | 1.000 | 1.000 | 1.000 |
| 2.0 | 0.977 | 0.979 | 0.981 | 0.986 | 0.987 | 0.988 | 0.990 |
| 3.0 | 0.905 | 0.925 | 0.938 | 0.942 | 0.949 | 0.951 | 0.954 |
| 4.0 | 0.861 | 0.881 | 0.896 | 0.902 | 0.909 | 0.913 | 0.916 |
| 5.0 | 0.818 | 0.840 | 0.851 | 0.860 | 0.868 | 0.874 | 0.879 |
| 6.0 | 0.777 | 0.797 | 0.810 | 0.818 | 0.829 | 0.836 | 0.840 |
| 7.0 | 0.753 | 0.762 | 0.770 | 0.781 | 0.790 | 0.798 | 0.803 |
| 8.0 | 0.716 | 0.724 | 0.733 | 0.745 | 0.752 | 0.761 | 0.767 |
| 9.0 | 0.680 | 0.688 | 0.699 | 0.710 | 0.716 | 0.727 | 0.730 |
| 10.0 | 0.646 | 0.655 | 0.665 | 0.675 | 0.685 | 0.693 | 0.698 |
| 12.0 | 0.584 | 0.593 | 0.600 | 0.611 | 0.619 | 0.627 | 0.634 |
| 14.0 | 0.527 | 0.536 | 0.544 | 0.554 | 0.561 | 0.569 | 0.575 |
| 16.0 | 0.476 | 0.484 | 0.493 | 0.502 | 0.508 | 0.517 | 0.521 |
| 18.0 | 0.430 | 0.437 | 0.447 | 0.455 | 0.460 | 0.469 | 0.472 |
| 20.0 | 0.388 | 0.395 | 0.405 | 0.412 | 0.417 | 0.427 | 0.428 |
| cGy/MU* | 0.900 | 0.910 | 0.920 | 0.930 | 0.940 | 0.940 | 0.950 |

* Output at 1.5-cm depth; normalized to 10×10 cm$^2$.

Adapted from Serago, C.F., Houdek, P.V., Hartmann, G.H., Saini, D.S., Serago, M.E., and Kaydee, A., Tissue maximum ratios (and other parameters) of small circular 4, 6, 10, 15 and 24 MV x-ray beams for radiosurgery, *Phys. Med. Biol.*, 37, 1943, 1992.

## TABLE 3.C3
### Tissue Phantom Ratios for 6-MV Stereotactic Beams [Varian Clinac-2500, 100-cm SAD]

| Depth (cm) | Field Diameter (cm) | | | | | | |
|---|---|---|---|---|---|---|---|
| | 1.25 | 1.5 | 2.0 | 2.5 | 3.0 | 3.25 | 3.5 |
| 1.5 | 1.000 | 1.000 | 1.000 | 1.000 | 1.000 | 1.000 | 1.000 |
| 2.0 | 0.963 | 0.966 | 0.971 | 0.974 | 0.976 | 0.976 | 0.977 |
| 3.0 | 0.924 | 0.930 | 0.940 | 0.945 | 0.950 | 0.950 | 0.952 |
| 4.0 | 0.885 | 0.893 | 0.906 | 0.913 | 0.919 | 0.920 | 0.922 |
| 5.0 | 0.846 | 0.855 | 0.869 | 0.878 | 0.886 | 0.887 | 0.889 |
| 6.0 | 0.808 | 0.818 | 0.832 | 0.842 | 0.851 | 0.853 | 0.854 |
| 7.0 | 0.771 | 0.780 | 0.794 | 0.805 | 0.815 | 0.817 | 0.818 |
| 8.0 | 0.736 | 0.744 | 0.757 | 0.769 | 0.778 | 0.781 | 0.782 |
| 9.0 | 0.701 | 0.709 | 0.721 | 0.733 | 0.743 | 0.746 | 0.746 |
| 10.0 | 0.669 | 0.676 | 0.686 | 0.699 | 0.709 | 0.712 | 0.712 |
| 12.0 | 0.609 | 0.615 | 0.624 | 0.637 | 0.646 | 0.649 | 0.648 |
| 14.0 | 0.556 | 0.561 | 0.570 | 0.582 | 0.591 | 0.594 | 0.593 |
| 16.0 | 0.507 | 0.512 | 0.523 | 0.533 | 0.543 | 0.546 | 0.545 |
| 18.0 | .0462 | 0.467 | 0.478 | 0.488 | 0.497 | 0.500 | 0.499 |
| 20.0 | 0.417 | 0.422 | 0.430 | 0.440 | 0.448 | 0.451 | 0.449 |
| cGy/MU* | 0.883 | 0.899 | 0.928 | 0.935 | 0.941 | 0.943 | 0.945 |

* Output at 1.5-cm depth; normalized to 10×10 cm$^2$ field.

Adapted from Neeranjun, W.S., Pennington, E.C., The University of Iowa, personal communication, 1992.

## TABLE 3.C4
## Tissue Phantom Ratios for 6-MV Stereotactic Beams [Philips SL75/5, 100-cm SAD]

| Depth (cm) | Field Diameter (cm) | | | | | | |
|---|---|---|---|---|---|---|---|
| | 1.0 | 1.5 | 2.0 | 2.5 | 3.0 | 3.5 | 4.0 |
| 1.0 | 0.993 | 0.982 | 0.975 | 0.974 | 0.971 | 0.969 | 0.967 |
| 1.5 | 1.000 | 1.000 | 1.000 | 1.000 | 1.000 | 1.000 | 1.000 |
| 2.0 | 0.982 | 0.989 | 0.991 | 0.993 | 0.995 | 0.997 | 0.998 |
| 2.5 | 0.959 | 0.968 | 0.973 | 0.976 | 0.978 | 0.981 | 0.984 |
| 3.0 | 0.937 | 0.946 | 0.954 | 0.959 | 0.962 | 0.966 | 0.970 |
| 4.0 | 0.895 | 0.904 | 0.913 | 0.920 | 0.926 | 0.930 | 0.936 |
| 5.0 | 0.855 | 0.866 | 0.874 | 0.881 | 0.888 | 0.894 | 0.899 |
| 6.0 | 0.815 | 0.826 | 0.836 | 0.844 | 0.852 | 0.859 | 0.866 |
| 7.0 | 0.778 | 0.789 | 0.798 | 0.805 | 0.818 | 0.825 | 0.832 |
| 8.0 | 0.743 | 0.754 | 0.762 | 0.770 | 0.781 | 0.789 | 0.796 |
| 9.0 | 0.710 | 0.721 | 0.729 | 0.738 | 0.749 | 0.756 | 0.763 |
| 10.0 | 0.677 | 0.689 | 0.696 | 0.706 | 0.716 | 0.725 | 0.732 |
| 12.0 | 0.614 | 0.627 | 0.636 | 0.646 | 0.657 | 0.664 | 0.670 |
| 14.0 | 0.562 | 0.574 | 0.581 | 0.590 | 0.600 | 0.605 | 0.601 |
| 16.0 | 0.514 | 0.522 | 0.531 | 0.538 | 0.550 | 0.554 | 0.559 |
| 18.0 | 0.471 | 0.477 | 0.487 | 0.493 | 0.502 | 0.508 | 0.514 |
| 20.0 | 0.428 | 0.437 | 0.446 | 0.453 | 0.460 | 0.466 | 0.472 |

Adapted from Bova, F., The University of Florida, Gainesville, FL, personal communication, 1992.

## TABLE 3.C5
### Tissue Phantom Ratios for 8-MV Stereotactic Beams [Philips SL75/20, 100-cm SAD]

| Depth (cm) | Field Diameter (cm) | | | | | | |
|---|---|---|---|---|---|---|---|
| | 1.0 | 1.5 | 2.0 | 2.5 | 3.0 | 3.5 | 4.0 |
| 1.0 | 0.960 | 0.938 | 0.917 | 0.910 | 0.904 | 0.902 | 0.901 |
| 1.5 | 0.996 | 0.989 | 0.980 | 0.975 | 0.970 | 0.975 | 0.980 |
| 2.0 | 1.000 | 1.000 | 1.000 | 1.000 | 1.000 | 1.000 | 1.000 |
| 3.0 | 0.961 | 0.970 | 0.980 | 0.983 | 0.987 | 0.990 | 0.994 |
| 4.0 | 0.923 | 0.935 | 0.948 | 0.952 | 0.956 | 0.961 | 0.967 |
| 5.0 | 0.884 | 0.890 | 0.910 | 0.917 | 0.925 | 0.929 | 0.934 |
| 6.0 | 0.851 | 0.863 | 0.876 | 0.884 | 0.893 | 0.898 | 0.904 |
| 7.0 | 0.818 | 0.831 | 0.844 | 0.852 | 0.861 | 0.866 | 0.872 |
| 8.0 | 0.784 | 0.798 | 0.813 | 0.821 | 0.830 | 0.835 | 0.841 |
| 9.0 | 0.751 | 0.766 | 0.782 | 0.790 | 0.799 | 0.805 | 0.811 |
| 10.0 | 0.723 | 0.740 | 0.751 | 0.760 | 0.770 | 0.776 | 0.783 |
| 12.0 | 0.668 | 0.681 | 0.695 | 0.704 | 0.714 | 0.720 | 0.726 |
| 14.0 | 0.618 | 0.630 | 0.644 | 0.652 | 0.661 | 0.668 | 0.675 |
| 16.0 | 0.570 | 0.580 | 0.598 | 0.605 | 0.612 | 0.619 | 0.627 |
| 18.0 | 0.526 | 0.540 | 0.554 | 0.560 | 0.567 | 0.574 | 0.582 |
| 20.0 | 0.488 | 0.500 | 0.512 | 0.519 | 0.527 | 0.532 | 0.538 |

From Bova, F., The University of Florida, Gainesville, FL, personal communication, 1992. With permission.

## TABLE 3.C6
### Tissue Phantom Ratios for 10-MV Stereotactic Beams [Varian Clinac-18, 100-cm SAD]

| Depth (cm) | Field Diameter (cm) | | | | | | |
|---|---|---|---|---|---|---|---|
| | 1.0 | 1.5 | 2.0 | 2.5 | 3.0 | 3.5 | 4.0 |
| 0.5 | 0.779 | 0.732 | 0.704 | 0.688 | 0.682 | 0.682 | 0.685 |
| 1.0 | 0.950 | 0.925 | 0.904 | 0.890 | 0.885 | 0.884 | 0.886 |
| 2.0 | 1.000 | 1.000 | 1.000 | 0.998 | 0.996 | 0.994 | 0.991 |
| 3.0 | 0.972 | 0.981 | 0.987 | 0.900 | 0.993 | 0.995 | 0.996 |
| 4.0 | 0.936 | 0.948 | 0.956 | 0.961 | 0.965 | 0.968 | 0.970 |
| 5.0 | 0.902 | 0.915 | 0.924 | 0.930 | 0.936 | 0.939 | 0.943 |
| 6.0 | 0.868 | 0.881 | 0.890 | 0.897 | 0.903 | 0.908 | 0.912 |
| 7.0 | 0.836 | 0.848 | 0.857 | 0.866 | 0.872 | 0.877 | 0.882 |
| 8.0 | 0.803 | 0.816 | 0.827 | 0.835 | 0.842 | 0.847 | 0.852 |
| 9.0 | 0.773 | 0.786 | 0.795 | 0.805 | 0.812 | 0.818 | 0.825 |
| 10.0 | 0.743 | 0.757 | 0.768 | 0.776 | 0.783 | 0.790 | 0.796 |
| 12.0 | 0.685 | 0.698 | 0.709 | 0.718 | 0.727 | 0.734 | 0.742 |
| 14.0 | 0.635 | 0.650 | 0.660 | 0.670 | 0.677 | 0.684 | 0.690 |
| 16.0 | 0.589 | 0.603 | 0.615 | 0.623 | 0.630 | 0.636 | 0.642 |
| 18.0 | 0.545 | 0.557 | 0.568 | 0.576 | 0.583 | 0.590 | 0.596 |
| 20.0 | 0.505 | 0.517 | 0.528 | 0.536 | 0.543 | 0.550 | 0.555 |
| cGy/MU* | 0.740 | 0.820 | 0.860 | 0.890 | 0.910 | 0.920 | 0.920 |

*Output at 2.5-cm depth; normalized to 10×10 cm$^2$.

From Serago, C.F., Houdek, P.V., Hartmann, G.H., Saini, D.S., Serago, M.E., and Kaydee, A., Tissue maximum ratios (and other parameters) of small circular 4, 6, 10, 15 and 24 MV x-ray beams for radiosurgery, *Phys. Med. Biol.*, 37, 1943, 1992. With permission.

## TABLE 3.C7
### Tissue Phantom Ratios for 15-MV Stereotactic Beams [Siemens Mevatron-77, 100-cm SAD]

| Depth (cm) | Field Diameter (cm) | | | | | | |
|---|---|---|---|---|---|---|---|
| | 1.0 | 1.5 | 2.0 | 2.5 | 3.0 | 3.5 | 4.0 |
| 0.5 | 0.533 | 0.487 | 0.503 | 0.545 | 0.563 | 0.557 | 0.554 |
| 1.0 | 0.861 | 0.806 | 0.791 | 0.803 | 0.801 | 0.788 | 0.779 |
| 2.0 | 0.998 | 0.987 | 0.978 | 0.977 | 0.973 | 0.965 | 0.960 |
| 3.0 | 0.996 | 0.997 | 1.000 | 1.000 | 1.000 | 1.000 | 1.000 |
| 4.0 | 0.972 | 0.975 | 0.982 | 0.984 | 0.988 | 0.993 | 0.996 |
| 5.0 | 0.941 | 0.946 | 0.956 | 0.959 | 0.965 | 0.972 | 0.977 |
| 6.0 | 0.910 | 0.916 | 0.927 | 0.931 | 0.938 | 0.947 | 0.953 |
| 7.0 | 0.880 | 0.887 | 0.898 | 0.902 | 0.910 | 0.920 | 0.927 |
| 8.0 | 0.851 | 0.859 | 0.870 | 0.874 | 0.883 | 0.894 | 0.900 |
| 9.0 | 0.822 | 0.831 | 0.842 | 0.847 | 0.856 | 0.867 | 0.874 |
| 10.0 | 0.795 | 0.804 | 0.815 | 0.821 | 0.829 | 0.841 | 0.848 |
| 12.0 | 0.742 | 0.753 | 0.764 | 0.770 | 0.779 | 0.790 | 0.798 |
| 14.0 | 0.693 | 0.705 | 0.716 | 0.722 | 0.731 | 0.743 | 0.750 |
| 16.0 | 0.647 | 0.660 | 0.671 | 0.678 | 0.687 | 0.698 | 0.706 |
| 18.0 | 0.605 | 0.618 | 0.629 | 0.636 | 0.645 | 0.656 | 0.664 |
| 20.0 | 0.565 | 0.578 | 0.590 | 0.597 | 0.606 | 0.617 | 0.624 |
| cGy/MU* | 0.620 | 0.760 | 0.830 | 0.870 | 0.880 | 0.910 | 0.940 |

*Output at 3.0-cm depth; normalized to $10 \times 10$ cm$^2$.

From Serago, C.F., Houdek, P.V., Hartmann, G.H., Saini, D.S., Serago, M.E., and Kaydee, A., Tissue maximum ratios (and other parameters) of small circular 4, 6, 10, 15 and 24 MV x-ray beams for radiosurgery, *Phys. Med. Biol.*, 37, 1943, 1992. With permission.

## TABLE 3.C8
### Tissue Phantom Ratios for 24-MV Stereotactic Beams [Varian Clinac-2500, 100-cm SAD]

| Depth (cm) | Field Diameter (cm) | | | | | | |
|---|---|---|---|---|---|---|---|
| | 1.0 | 1.5 | 2.0 | 2.5 | 3.0 | 3.5 | 4.0 |
| 1.0 | 0.784 | 0.732 | 0.699 | 0.690 | 0.686 | 0.685 | 0.684 |
| 2.0 | 0.965 | 0.936 | 0.917 | 0.906 | 0.900 | 0.897 | 0.896 |
| 3.0 | 0.998 | 0.991 | 0.985 | 0.980 | 0.976 | 0.975 | 0.975 |
| 4.0 | 0.994 | 0.998 | 1.000 | 1.000 | 1.000 | 1.000 | 1.000 |
| 5.0 | 0.974 | 0.983 | 0.988 | 0.991 | 0.993 | 0.994 | 0.995 |
| 6.0 | 0.951 | 0.962 | 0.969 | 0.973 | 0.976 | 0.978 | 0.979 |
| 7.0 | 0.928 | 0.938 | 0.947 | 0.951 | 0.955 | 0.957 | 0.958 |
| 8.0 | 0.901 | 0.914 | 0.923 | 0.929 | 0.932 | 0.935 | 0.936 |
| 9.0 | 0.877 | 0.889 | 0.900 | 0.905 | 0.908 | 0.912 | 0.914 |
| 10.0 | 0.853 | 0.865 | 0.876 | 0.882 | 0.886 | 0.889 | 0.892 |
| 12.0 | 0.807 | 0.820 | 0.832 | 0.836 | 0.842 | 0.845 | 0.848 |
| 14.0 | 0.752 | 0.778 | 0.788 | 0.795 | 0.800 | 0.804 | 0.806 |
| 16.0 | 0.720 | 0.738 | 0.748 | 0.755 | 0.760 | 0.764 | 0.767 |
| 18.0 | 0.682 | 0.699 | 0.710 | 0.718 | 0.722 | 0.726 | 0.728 |
| 20.0 | 0.648 | 0.662 | 0.674 | 0.681 | 0.686 | 0.690 | 0.693 |
| cGy/MU* | 0.590 | 0.710 | 0.780 | 0.830 | 0.860 | 0.880 | 0.890 |

*Output at 4-cm depth; normalized to 10×10 cm$^2$.

From Serago, C.F., Houdek. P.V., Hartmann, G.H., Saini, D.S., Serago, M.E., and Kaydee, A., Tissue maximum ratios (and other parameters) of small circular 4, 6, 10, 15 and 24 MV x-ray beams for radiosurgery, *Phys. Med. Biol.*, 37, 1943, 1992. With permission.

## TABLE 3.D1
### Depth Dose Data for Flat IORT Cones [Varian Clinac-2500, 118-cm SSD]

| Electron Beam Energy | Depth $R$ at which Dose $D$ Occurs* $R_D$ (cm) | | | | | Surface Dose $D_s$ (%) | X Ray Dose $D_x$ (%) |
|---|---|---|---|---|---|---|---|
| | $R_{100}$ | $R_{90}$ | $R_{80}$ | $R_{50}$ | $R_p$ | | |
| 6 MeV | 0.8 | 1.3 | 1.5 | 1.8 | 2.5 | 85.0 | <1 |
| 9 MeV | 1.5 | 2.1 | 2.4 | 2.9 | 3.7 | 86.0 | <1 |
| 12 MeV | 2.0 | 3.0 | 3.4 | 4.2 | 5.3 | 89.0 | 1 |
| 15 MeV | 2.5 | 4.0 | 4.5 | 5.4 | 6.8 | 91.0 | 1 |
| 18 MeV | 2.6 | 4.7 | 5.4 | 6.7 | 8.4 | 92.0 | 2 |
| 22 MeV | 2.7 | 5.5 | 6.3 | 7.8 | 10.3 | 92.0 | 3 |

* For all flat cones except 4 cm size (see text). X ray collimator setting=10×10 cm².

Adapted from Jani, S.K., The University of Iowa Hospitals and Clinics, Iowa City, IA, personal communication, 1987.

## TABLE 3.D2
### Depth Dose Data for 15° Beveled IORT Cones [Varian Clinac-2500, 118-cm SSD]

| Electron Beam Energy | Depth $R$ at which Dose $D$ Occurs* $R_D$ (cm) | | | | | Surface Dose $D_s$ (%) | X Ray Dose $D_x$ (%) |
|---|---|---|---|---|---|---|---|
| | $R_{100}$ | $R_{90}$ | $R_{80}$ | $R_{50}$ | $R_p$ | | |
| 6 MeV | 0.7 | 1.2 | 1.3 | 1.7 | 2.4 | 87.0 | <1 |
| 9 MeV | 1.2 | 1.9 | 2.2 | 2.8 | 3.5 | 87.0 | <1 |
| 12 MeV | 1.9 | 2.9 | 3.2 | 4.0 | 5.2 | 87.0 | 1 |
| 15 MeV | 2.4 | 3.7 | 4.2 | 5.2 | 6.6 | 88.0 | 1 |
| 18 MeV | 2.7 | 4.5 | 5.1 | 6.3 | 8.2 | 89.0 | 2 |
| 22 MeV | 3.1 | 5.1 | 6.0 | 7.4 | 9.8 | 90.0 | 3 |

* For all 15° beveled cones except 4-cm size (see text). X ray collimator setting=10×10 cm².

Adapted from Jani, S.K., The University of Iowa Hospitals and Clinics, Iowa City, IA, personal communication, 1987.

TABLE 3.D3

Depth Dose Data for 30° Beveled IORT Cones [Varian Clinac-2500, 118-cm SSD]

| Electron Beam | | Depth R at which Dose D Occurs* $R_D$ (cm) | | | | | Surface Dose | X Ray Dose |
|---|---|---|---|---|---|---|---|---|
| Energy | $R_{100}$ | $R_{90}$ | $R_{80}$ | $R_{50}$ | $R_p$ | | $D_s$ (%) | $D_x$ (%) |
| 6 MeV | 0.5 | 0.9 | 1.1 | 1.5 | 2.3 | | 90.0 | <1 |
| 9 MeV | 0.9 | 1.6 | 1.8 | 2.5 | 3.4 | | 87.0 | <1 |
| 12 MeV | 1.4 | 2.3 | 2.7 | 3.6 | 5.0 | | 85.0 | 1 |
| 15 MeV | 1.9 | 2.8 | 3.4 | 4.2 | 6.2 | | 85.0 | 1 |
| 18 MeV | 2.3 | 3.5 | 4.1 | 4.7 | 6.8 | | 85.0 | 2 |
| 22 MeV | 2.6 | 4.1 | 4.6 | 5.2 | 7.1 | | 85.0 | 3 |

* For all 30° beveled cones except 4 cm size (see text). X ray collimator setting=10×10 cm².

Adapted from Jani, S.K., The University of Iowa Hospitals and Clinics, Iowa City, IA, personal communication, 1987.

TABLE 3.D4

Electron Beam Output for Flat IORT Cones [Varian Clinac-2500, 118 cm SSD]

| Electron Beam Energy | Depth of Measurement (cm) | Output [ cGy/MU]* | | | | | | |
|---|---|---|---|---|---|---|---|---|
| | | Cone Diameter (cm) | | | | | | |
| | | 4 | 5 | 6 | 6.5 | 7 | 7.5 | 8 | 9 |
| 6 MeV | 1.0 | 0.287 | 0.355 | 0.381 | 0.394 | 0.404 | 0.411 | 0.420 | 0.429 |
| 9 MeV | 1.5 | 0.333 | 0.383 | 0.396 | 0.402 | 0.404 | 0.406 | 0.413 | 0.416 |
| 12 MeV | 2.0 | 0.449 | 0.492 | 0.500 | 0.502 | 0.504 | 0.505 | 0.507 | 0.509 |
| 15 MeV | 2.5 | 0.610 | 0.650 | 0.651 | 0.652 | 0.653 | 0.654 | 0.655 | 0.656 |
| 18 MeV | 2.5 | 0.685 | 0.704 | 0.702 | 0.700 | 0.699 | 0.698 | 0.700 | 0.700 |
| 22 MeV | 2.5 | 0.714 | 0.722 | 0.713 | 0.709 | 0.708 | 0.707 | 0.709 | 0.712 |

* X ray collimator setting=10×10cm²; standard 10×10 cm² cone ooutput=1.00cGy/MU

Adapted from Jani, S.K., The University of Iowa Hospitals and Clinics, Iowa City, IA, personal communication, 1987.

## TABLE 3.D5
### Electron Beam Output for 15° Beveled IORT Cones [Varian Clinac-2500, 118-cm SSD]

| Electron Beam Energy | Depth of Measurement (cm) | Output [ cGy/MU]* | | | | | | | |
|---|---|---|---|---|---|---|---|---|---|
| | | Cone Diameter (cm) | | | | | | | |
| | | 4 | 5 | 6 | 6.5 | 7 | 7.5 | 8 | 9 |
| 6 MeV | 1.0 | 0.302 | 0.386 | 0.409 | 0.427 | 0.436 | 0.442 | 0.454 | 0.459 |
| 9 MeV | 1.5 | 0.326 | 0.388 | 0.400 | 0.408 | 0.409 | 0.412 | 0.417 | 0.417 |
| 12 MeV | 2.0 | 0.437 | 0.492 | 0.500 | 0.504 | 0.505 | 0.504 | 0.507 | 0.509 |
| 15 MeV | 2.5 | 0.591 | 0.646 | 0.650 | 0.652 | 0.650 | 0.647 | 0.650 | 0.652 |
| 18 MeV | 2.5 | 0.669 | 0.698 | 0.698 | 0.697 | 0.695 | 0.696 | 0.695 | 0.696 |
| 22 MeV | 3.0 | 0.692 | 0.714 | 0.713 | 0.711 | 0.708 | 0.706 | 0.710 | 0.713 |

* X ray collimator setting=10×10cm$^2$; standard 10×10 cm$^2$ cone output =1.00 cGy/MU.

Adapted from Jani, S.K., The University of Iowa Hospitals and Clinics, Iowa City, IA, personal communication, 1987.

## TABLE 3.D6
### Electron Beam Output for 30° Beveled IORT Cones [Varian Clinac-2500, 118-cm SSD]

| Electron Beam Energy | Depth of Measurement (cm) | Output [ cGy/MU]* | | | | | | | |
|---|---|---|---|---|---|---|---|---|---|
| | | Cone Diameter (cm) | | | | | | | |
| | | 4 | 5 | 6 | 6.5 | 7 | 7.5 | 8 | 9 |
| 6 MeV | — | — | — | — | — | — | — | — | — |
| 9 MeV | 1.0 | 0.333 | 0.388 | 0.401 | 0.413 | 0.414 | 0.419 | 0.422 | 0.424 |
| 12 MeV | 1.5 | 0.452 | 0.501 | 0.508 | 0.516 | 0.517 | 0.524 | 0.524 | 0.524 |
| 15 MeV | 2.0 | 0.607 | 0.660 | 0.662 | 0.663 | 0.665 | 0.667 | 0.670 | 0.672 |
| 18 MeV | 2.5 | 0.645 | 0.700 | 0.702 | 0.704 | 0.706 | 0.708 | 0.710 | 0.712 |
| 22 MeV | 2.5 | 0.687 | 0.720 | 0.718 | 0.717 | 0.714 | 0.716 | 0.716 | 0.717 |

* X ray collimator setting=10×10cm$^2$; standard 10×10 cm$^2$ cone output =1.00 cGy/MU.

Adapted from Jani, S.K., The University of Iowa Hospitals and Clinics, Iowa City, IA, personal communication, 1987.

# REFERENCES

## A. TOTAL BODY PHOTON IRRADIATION

1. **AAPM**, The Physical Aspects of Total and Half Body Photon Irradiation, American Association of Physicists in Medicine, AAPM Report No. 17, American Institute of Physics, New York, 1986.

2. **Aget, H., Van Dyk, J., and Leung, P.M.K.** Utilization of a high energy photon beam for whole body irradiation, *Radiol.*, 123, 747, 1977.

3. **Christ, G.**, The dosimetry of total body irradiation at the University of Tubingen, *J. Eur. Radiother.*, 3, 211, 1982.

4. Conference on dosimetry of total body irradiation by external photon beams, Cloutier, R.J., O'Foghludha, F. and Comas, F.V. Eds., Document CONF-670219, Health and Safety (TID-4500) Oak Ridge Associated Universities, 1967.

5. **Cunningham, J.R. and Wright, D.J.**, A simple facility for whole-body irradiation, *Radiol.*, 78, 941, 1962.

6. **Curran, W.J., Galvin, J.M., and D'Angio, G.J.**, A simple dose calculation method for total body photon irradation, *Int. J. Radiat. Oncol. Biol. Phys.*, 17, 219, 1989.

7. **Engler, M.J., Feldman, M.I., and Spira, J.**, Arc technique for total body irradiation by a 42 MV betatron, *Med. Phys.*, 4, 524, 1977.

8. **Findley, D.O., Skov, D.D., and Blume, K.G.**, Total body irradiation with a 10 MV linear accelerator in conjunction with bone marrow transplantation, *Int. J. Radiat. Oncol. Biol. Phys.*, 6, 695, 1980.

9. **Fitzpatrick P.J. and Rider, W.D.**, Half body radiotherapy, *Int. J. Radiat. Oncol. Biol. Phys.*, 1, 197, 1976.

10. **Galvin, J.M.**, Calculation and prescription of dose for total body irradiation, *Int. J. Radiat. Oncol. Biol. Phys.*, 9, 1919, 1983.

11. **Galvin, J.M., D'Angio, G.J., and Walsh, G.**, Use of tissue compensators to improve the dose uniformity for total body irradiation, *Int. J. Radiat. Oncol. Biol. Phys.*, 6, 767, 1980.

12. **Glasgow, G.P. and Mill, W.B.**, Cobalt-60 total body irradiation dosimetry at 220 cm source-axis distance, *Int. J. Radiat. Oncol. Biol. Phys.*, 6, 773, 1980.

13. **Glasgow, G.P., Mill, W.B., Phillips, G.L., and Herzig, G.P.**, Comparative 60-Co total body irradiation (220 cm SAD) and 25 MV total body irradiation (370 cm SAD) dosimetry, *Int. J. Radiat. Oncol. Biol. Phys.*, 6, 1243, 1980.

14. **Glasgow, G.P.**, The dosimetry of fixed, single source hemibody and total body irradiators, *Med. Phys.*, 9, 311, 1982.

15. **Houdek, P.V. and Pisciotti, V.J.**, A comparison of calculated and measured data for total body irradiation by 10 MV x-rays, *Phys. Med. Biol.*, 32, 1101, 1987.

16. **Jani, S.K. and Pennington, E.C.**, University of Iowa, personal communication, 1992.

17. **Jani, S.K. and Pennington, E.C.**, Depth dose characteristics of 24 MV x-ray beams at extended SDD, *Med. Phys.*, 18, 292, 1991.

18. **Jani, S.K., Pennington, E.C. and Wen, B.C.**, Dose buildup characteristics of a cobalt-60 TBI field, *Med. Phys.*, 514, 1990.

19. **Khan, F.M., Williamson, J.F., Sewchand W., and Kim, T.H.**, Basic data for dosage calculation and compensation, *Int. J. Radiat. Oncol. Biol. Phys.*, 6, 745, 1980.

20. **Kim, T.H., Khan, F.M., and Galvin, J.M.**, A report of the work party: comparison of total body irradiation techniques for bone marrow transplantation, *Int. J. Radiat. Oncol. Biol. Phys.*, 6, 779, 1980.

21. **Lam, W.C., Lindskoug, B.A., Order, S.E., and Grant, D.G.**, The dosimetry of cobalt-60 total body irradiation, *Int. J. Radiat. Oncol. Biol. Phys.*, 5, 905, 1979.

22. **Lam, W.C., Order, S.E., and Thomas, E.D.**, Uniformity and standardization of single and opposing cobalt-60 sources for total body irradiation, *Int. J. Radiat. Oncol. Biol. Phys.*, 6, 245, 1980.

23. **Leung, P.M.K., Rider, W.D., Webb, H.P., Aget, H., and Johns, H.E.**, Cobalt-60 therapy unit for large field irradiation, *Int. J. Radiat. Oncol. Biol. Phys.*, 7, 705, 1981.

24. **Lutz, W.R., Dougan, P.W., and Bjarngard, B.E.**, Design and characteristics of a facility for total body and large field irradiation, *Int. J. Radiat. Oncol. Biol. Phys.*, 15, 1035, 1988.

25. **Marinello, G., Barrie, A.M., and LeBourgeois, J.P.**, Measurement and calculation of lung dose in total body irradiation performed with cobalt-60, *J. Eur. Radiotherap.*, 3, 174, 1982.

26. **Miller, R.J., Langdon, E.A., and Tesler, A.S.**, Total body irradiation utilizing a single cobalt-60 source, *Int. J. Radiat. Oncol. Biol. Phys.,* 1, 549, 1976.

27. **Mulvey, P.J. and Godlee, J.N.**, Technique and dosimetry for TBI at University of College Hospital, London, *J. Eur. Radiother.*, 3, 241, 1982.

28. **Niroomand-Rad, A.**, Physical aspects of total body irradiation of bone marrow transplant patients using 18 MV x rays, *Int. J. Radiat. Oncol. Biol. Phys.,* 20, 605, 1991.

29. **Obcemea, C.H., Rice, R.K., Mijnheer, B.J., Siddon, R.L., Tarbell, N.J., Mauch, P., and Chin, L.M.**, Three-dimensional dose distribution of total body irradiation by a dual source total body irradiator, *Int. J. Radiat. Oncol. Biol. Phys.,* 24, 789, 1992.

30. **Peters, V.G. and Herer, A.S.**, Modification of a standard cobalt-60 unit for total body irradiation at 150 cm SSD, *Int. J. Radiat. Oncol. Biol. Phys.*, 10, 927, 1984.

31. **Pla, M., Chenery, S.G., and Podgorsak, E.B.**, Total body irradiation with a sweeping beam, *Int. J. Radiat. Oncol. Biol. Phys.*, 9, 83, 1983.

32. **Podgorsak, E.B., Pla, C., Evans, M.D.C., and Pla, M.**, The influence of phantom size on output, peak-scatter factor and percentage depth dose in large field photon irradiation, *Med. Phys.*, 12, 639, 1985.

33. **Quast, U.**, Physical treatment planning of total body irradiation — patient translation and beam zone method, *Med. Phys.*, 12, 567, 1985.

34. **Rider, W.D. and Van Dyk, J.**, Total and partial body irradiation, In *Radiation Therapy Treatment Planning*, Glatstein E. and Haybittle, J.L., Eds., Marcel Dekker Inc., Bleehen, N.M., 1983, 559.

35.  **Shank, B.**, Techniques of magna-field irradiation, *Int. J. Radiat. Oncol. Biol. Phys.*, 9, 1925, 1983.

36.  **Svensson, G.K., Larson, R.D., and Chen, T.S.**, The use of a 4 MV linear accelerator for whole body irradiation, *Int. J. Radiat. Oncol. Biol. Phys.*, 6, 761, 1980.

37.  The physics and clinical aspects of total body irradiation used for bone marrow transplantation, Workshop of the Children's Cancer Study Group, Montreal, June, 1978, *Int. J. Radiat. Oncol. Biol. Phys.*, 6, 743, 1980.

38.  **Van Dyk, J.**, Advances in radiation therapy, in *Whole and Partial Body Radiotherapy: Physical Considerations,* AAPM, American Association of Physicists in Medicine, AAPM Monograph No. 9, Wright, A. and Boyer, A., Eds., American Institute of Physics, New York, 1986, 403.

39.  **Van Dyk, J., Battista, J.J., and Rider, W.D.**, Half body radiotherapy: The use of computed tomography to determine the dose to lung, *Int. J. Radiat. Oncol. Biol. Phys.*, 6, 463, 1980.

40.  **Van Dyk, J.**, Broad beam attenuation of cobalt-60 gamma rays and 6, 18 and 25 MV x rays by lead, *Med. Phys.*, 13, 105, 1986.

41.  **Van Dyk, J., Keane, T.J., and Rider, W.D.**, Lung density as measured by computerized tomography: implications for radiotherapy, *Int. J. Radiat. Oncol. Biol. Phys.*, 8, 1363, 1982.

42.  **Van Dyk, J., Leung, P.M.K., and Cunningham, J.R.**, Dosimetric considerations of very large cobalt-60 fields, *Int. J. Radiat. Oncol. Biol. Phys.*, 6, 753, 1980.

43.  **Van Dyk, J.**, Magna-field irradiation: physical considerations, *Int. J. Radiat. Oncol. Biol. Phys.*, 9, 1913, 1983.

44.  **Van Dyk, J.**, Whole and partial body radiotherapy: physical considerations, in Advances in Radiation Therapy Treatment Planning, Wright A. and Boyer, A.L., Eds., AAPM Monograph No. 9, American Institute of Physics, N.Y., 1983, 403.

## B.  TOTAL BODY ELECTRON THERAPY

45.  **AAPM**, American Association of Physicists in Medicine, Total Skin Electron Therapy: Technique and Dosimetry, AAPM Report No. 23 American Institute of Physics, New York, 1988.

46.  **Asbell, S.O., Siu, J., Lightfoot, D.A., and Brady, L.W.**, Individualized eye shields for use in electron beam therapy as well as low-energy photon irradiation, *Int. J. Rad. Oncol. Biol. Phys.*, 6, 519, 1980.

47.  **Biggs, P.J.**, The effect of beam angulation on central axis percent depth dose for 4–29 MeV electrons, *Phys. Med. Biol.*, 29, 1089, 1984.

48.  **Bjarngard, B.E., Chen, G.T.Y., Piontek, R.W., and Svensson, G.K.**, Analysis of dose distributions in whole body superficial electron therapy, *Int. J. Radiat. Oncol. Biol. Phys.*, 2, 319, 1977.

49.  **Buechner, W.W., Van de Graaff, R.J., Burrill, E.A., and Sperduto, A.**, Thick-target x-ray production in the range from 1250–2350 kilovolts, *Phys. Rev.*, 74, 1348, 1948.

50. **Chu, F.C.H. and Laughlin, J.S.,** Total skin electron beam therapy, In *Proceedings of the Symposium on Electron Beam Therapy*, New York, 1979.

51. **Coffey, C.W., Maruyama, Y., Stewart, B.L., and White, G.A.,** Electron beam irradiation for mycosis fungoides using variable energy, *J. of Kentucky Med. Assoc.*, 80, 398, 1982.

52. **Cox, R.S., Heck, R.J., Fessenden, P., Karzmark, C.J., and Rust, D.C.,** Development of total-skin electron therapy at two energies, *Int. J. Radiat. Oncol. Biol. Phys.,* 18, 659, 1990.

53. **Das, I.J., Kase, K.R., Copeland, J.F., and Fitzgerald, T.J.,** Electron beam modifications for the treatment of superficial malignancies, *Int. J. Radiat. Oncol. Biol. Phys.*, 21, 1627, 1991.

54. **Edelstein, G.R., Clark, T., and Holt, J.G.,** Dosimetry for total-body electron-beam therapy in the treatment of mycosis fungoides, *Radiology* 108, 691, 1973.

55. **Ekstrand, K.E. and Dixon, R.L.,** The problem of obliquely incident beams in electron-beam treatment planning, *Med. Phys.*, 9, 276, 1982.

56. **Fraass, B.A., Roberson, P.L., and Glatstein, E.,** Whole-skin electron treatment: Patient skin dose distribution, *Radiol.*, 146, 811, 1983.

57. **Galbraith, D.M. and Rawlinson, J.A.,** Partial bolussing to improve the depth doses in the surface region of low energy electron beams, *Int. J. Radiat. Oncol. Biol. Phys.*, 10, 313, 1984.

58. **Grollman, Jr., J.H., Bierman, S.M., Morgan, J.E., and Ottoman, R.E.,** X-ray contamination in total-skin electron therapy of lymphoma cutis and exfoliative dermatitis, *Radiol.*, 85, 356, 1965.

59. **Holt, J.G. and Perry, D.J.,** Some physical considerations in whole skin electron beam therapy, *Med. Phys.*, 9, 769, 1982.

60. **Kao, M., Lanzl, L.H., Rozenfeld, M., Kramer, T., and Chung-Bin, A.,** Electron whole body treatment dose analysis of Stanford technique, *Med. Phys.*, 11, 379, 1984 (abstract).

61. **Karzmark, C.J., Loevinger, R., Steele, R.E., Weissbluth, M.,** A technique for large field, superficial electron therapy, *Radiol.*, 74, 633, 1960.

62. **Kim, T.H., Pla, C., Pla, M., and Podgorsak, E.B.,** Clinical aspects of a rotational total skin electron irradiation, *Br. J. Radiol.*, 57, 501, 1984.

63. **Kitagawa, T.,** 10 MeV betatron electron beam therapy adapted to a case of mycosis fungoides, *Am. J. Roent. Rad. Therapy and Nuc. Med.*, 88, 229, 1962.

64. **Kumar, P.P. and Patel, I.S.,** Comparison of dose distribution with different techniques of total skin electron beam therapy, *Clinical Radiol.*, 33, 495, 1982.

65. **Kumar, P.P. and Patel, I.S.,** Rotation technique for superficial total body electron beam irradiation, *J. Natl. Med. Assoc.*, 70, 507, 1978.

66. **Kumar, P.P., Henschke, U.K., and Nibhanupudy, J.R.,** Problems and solutions in achieving uniform dose distribution in superficial total body electron therapy, *J. Natl. Med. Assoc.*, 69, 645, 1977.

67. **Kumar, P.P., Henschke, U.K., Mandal, K.P., Nibhanupudy, J.R., and Patel, I.S.,** Early experience in using an 18 MeV linear accelerator for

mycosis fungoides at Howard University Hospital, *J. of Natl. Med. Assoc.*, 69, 223, 1977.

68. **Leavitt, D.D., Stewart, J.R., and Earley, L.**, Improved dose homogeneity in electron arc therapy achieved by a multiple-energy technique, *Int. J. Radiat. Oncol. Biol. Phys.*, 19, 159, 1990.

69. **Leavitt, D.D., Stewart, J.R., Moeller, J.H., and Earley, L.**, Optimization of electron arc therapy doses by multi-vane collimator control, *Int. J. Radiat. Oncol. Biol. Phys.*, 16, 489, 1988.

70. **Lo, T.C.M., Salzman, F.A., Moschella, S.L., Tolman, E.L., and Wright, K.A.**, Whole body surface electron irradiation in the treatment of mycosis fungoides, *Radiol.* 130, 453, 1979.

71. **Meyler, T.S., Blumberg, A.L., and Purser, P.**, Total skin electron beam therapy, *Cancer*, 42, 1171, 1978.

72. **Niroomand-Rad, A., Gillin, M.T., Komaki, R., Kline, R.W., and Grimm, D.F.**, Dose distribution in total skin electron beam irradiation using the six-field technique, *Int. J. Rad. Oncol. Biol. Phys.*, 12, 415, 1986.

73. **Page, V., Gardner, A., and Karzmark, C.J.**, Patient dosimetry in the electron treatment of large superficial lesions, *Radiol.*, 94, 635, 1970.

74. **Pennington, E.C.**, The University of Iowa, personal communication, 1992.

75. **Pla, C., Heese, R., Pla, M., and Podgorsak, E.B.**, Calculation of surface dose in rotational total skin electron irradiation, *Med. Phys.*, 11, 539, 1984.

76. **Podgorsak, E.B., Pla, C., Pla, M., Lefebvre, P.Y., and Heese, R.**, Physical aspects of a rotational total skin electron irradiation, *Med. Phys.*, 10, 159, 1983.

77. **Sewchand, W., Khan, F.M., and Williamson, J.**, Total-body superficial electron-beam therapy using a multiple-field pendulum-arc technique, *Radiol.*, 130, 493, 1979.

78. **Sharma, S.C. and Wilson, D.L.**, Dosimetric study of total skin irradiation with a scanning beam electron accelerator, *Med. Phys.*, 14, 355, 1987.

79. **Tetenes, P.J. and Goodwin, P.N.**, Comparative study of superficial whole body radiotherapeutic techniques using a 4 MeV nonangulated electron beam, *Radiology*, 122, 219, 1977.

80. **Williams, P.C., Hunter, R.D., Jackson, S.M.**, Whole body electron therapy in mycosis fungoides—a successful translational technique achieved by modification of an established linear accelerator, *Br. J. Radiol.*, 52, 302, 1979.

## C. STEREOTACTIC RADIOSURGERY

81. **Arias, S., Schwade, J.G., Avitbol, A., and Marcial-Vega, Serago, C.F., Lewin, A.A., Houdek, P.V., and Gonzales, V.**, Radiosurgery target point alignment errors detected with portal film verification, *Int. J. Radiat. Oncol. Biol. Phys.*, 24, 777, 1992.

82. **Bjärngard, B.E., Tsai, J.S., and Rice, R.K.**, Doses on the central axes of narrow 6-MV x-ray beams, *Med. Phys.*, 17, 794, 1990.

83. **Bova, F.**, The University of Florida, Gainesville, FL, personal communication, 1992.

84. **Brenner, D.J., Martel, M.K., and Hall, E.J.,** Fractionated regimens for stereotactic radiotherapy of recurrent tumors in the brain, *Int. J. Radiat. Oncol. Biol. Phys.,* 21, 819, 1991.

85. **Bianciardi, L., D'Angelo, L., Gentile, F.P., Benassi, M., and Guerra, A.S.,** Dosimetry of small x-ray radiation fields, in Dosimetry in Radiotherapy, IAEA Rep. SM-298/42, Vienna, 1988, 355.

86. **Coffey, R.J., Flickinger, J.C., Bissonette, D.J., and Lunsford, L.D.,** Radiosurgery for solitary brain metastases using the cobalt-60 gamma unit: methods and results in 24 patients, *Int. J. Radiat. Oncol. Biol. Phys.,* 20, 1287, 1991.

87. **Colombo, F., Benedetti A., Pozza, F., Zanardo, A., Avanzo, R.C., Chierego, G., and Marchetti, C.,** Stereotactic radiosurgery utilizing a linear accelerator, *Appl. Neurophysiol.,* 48, 133, 1985.

88. **Colombo, F., et al.,** Radiosurgery using a 4 MV linear accelerator, *Acta Radiol. Suppl.,* 369, 603, 1986.

89. **Colombo, F., Benedetti, A., Possa, F., Avanzo, R.C., Marchetti, C., Chierego, G., and Zanardo, A.,** External stereotactic irradiation by linear accelerator, *Neurosurgery,* 16, 154, 1985.

90. **Delannes, M., Daly, N.J., Bonnet, J., Sabatier, J., and Tremoulet, M.,** Fractionated radiotherapy of small inoperable lesions of the brain using a non-invasive stereotactic frame, *Int. J. Radiat. Oncol. Biol. Phys.,* 21, 749, 1991.

91. **Eden, B.V. and Larner, J.M.,** The role of radiosurgery in arteriovenous malformations, *Applied Radiol.,* 71, June, 1991.

92. **Engenhart, R., Kimmig, B.N., Höver, K.H., Wowra, B., Sturm, V., van Kaick, G., and Wannenmacher, M.,** Stereotactic single high dose radiation therapy of benign intracranial meningiomas, *Int. J. Radiat. Oncol. Biol. Phys.,* 19, 1021, 1990.

93. **Flickinger, J.C. and Steiner, L.,** Radiosurgery and the double logistic product formula, *Radiotherap. and Oncol.,* 17, 229, 1990.

94. **Flickinger, J.C., Lunsford, L.D., Wu, A., Maitz, A.H., Kalend, A.M.,** Treatment planning for gamma knife radiosurgery with multiple isocenters, *Int. J. Radiat. Oncol. Biol. Phys.,* 18, 1495, 1990.

95. **Flickinger, J.C., Maitz, A., Kalend, A., Lunsford, L.D., and Wu, A.,** Treatment volume shaping with selective beam blocking using the Leksell gamma unit, *Int. J. Radiat. Oncol. Biol. Phys.,* 19, 783, 1990.

96. **Flickinger, J.C., Schell, M.C., and Larson, D.A.,** Estimation of complications for linear accelerator radiosurgery with the integrated logistic formula, *Int. J. Radiat. Oncol. Biol. Phys.,* 19, 143, 1990.

97. **Friedman, W.A. and Bova, F.J.,** The University of Florida radiosurgery system, *Surg. Neurol.,* 32, 334, 1989.

98. **Gehring, M.A., Mackie, T.R., Kubsad, S.S., Paliwal, B.R., Mehta, M.P., and Kinsella, T.J.,** A three-dimensional volume visualization package applied to stereotactic radiosurgery treatment planning, *Int. J. Radiat. Oncol. Biol. Phys.,* 21, 491, 1991.

99. **Gill, S.S., Thomas, D.G.T., Warrington, A.P., and Brada, M.,** Relocatable frame for stereotactic external beam radiotherapy, *Int. J. Radiat. Oncol. Biol. Phys.,* 20, 599, 1991.

100.  **Graham, J.D., Nahum, A.E., and Brada, M.**, A comparison of techniques for stereotactic radiotherapy by linear accelerator based on 3-dimensional dose distributions, *Radiotherap. and Oncol.*, 22, 29, 1991.

101.  **Graham, J.D., Warrington, A.P., Gill, S.S., and Brada, M.**, A non-invasive, relocatable stereotactic frame for fractionated radiotherapy and multiple imaging, *Radiotherap. and Oncol.*, 21, 60, 1991.

102.  **Hariz, M.I., Henriksson R., Löfroth, P-O, Laitinen, L.V., and Säterborg, N-E.**, A non-invasive method for fractionated stereotactic irradiation of brain tumors with linear accelerator, *Radiotherap. and Oncol.*, 17, 57, 1990.

103.  **Houdek, P.,V., VanBuren, J.M., and Fayos, J.V.**, Dosimetry of small radiation fields for 10-MV X rays, *Med. Phys.*, 10, 333, 1983.

104.  **Houdek, P.V., Fayos, J.V., Van Buren, J.M., and Ginsberg, M.S.**, Stereotaxic radiotherapy technique for small intracranial lesions, *Med. Phys.*, 12, 469, 1985.

105.  **Kubsad, S.S., Mackie, T.R., Gehring, M.A., Misisco, D.J., Paliwal, B.R., Mehta, M.P., and Kinsella, T.J.**, Monte carlo and convolution dosimetry for stereotactic radiosurgery, *Int. J. Radiat. Oncol. Biol. Phys.*, 19, 1027, 1990.

106.  **Larson, D.A., Gutin, P.H., Leibel, S.A., Phillips, T.L., Sneed, P.K., and Wara, W.M.**, Stereotaxic irradiation of brain tumors, *Cancer (Suppl)* 65, 792, 1990.

107.  **Leavitt, D.D., Gibbs Jr., F.A., Heilbrun, M.P., Moeller, J.H., and Takach Jr., G.A.**, Dynamic field shaping to optimize stereotactic radiosurgery, *Int. J. Radiat. Oncol. Biol. Phys.*, 21, 1247, 1991.

108.  **Leksell, D.G.**, Stereotactic radiosurgery, *Neurol. Res.*, 9, 60, 1987.

109.  **Leksell, L.**, The stereotaxic method and radiosurgery of the brain, *Acta Chir. Scand.*, 102, 316, 1951.

110.  **Levy, R.P., Fabrikant, J.I., Frankel, K.A., Phillips, M.H., and Lyman, J.T.**, Stereotactic heavy-charged-particle Bragg peak radiosurgery for the treatment of intracranial arteriovenous malformations in childhood and adolescence, *Neurosurgery*, 24, 841, 1989.

111.  **Lunsford, L.D., Flickinger, J., Lindner, G., and Maitz, A.**, Stereotactic radiosurgery of the brain using the first United States 201 Cobalt-60 source gamma knife, *Neurosurgery*, 24, 151, 1989.

112.  **Lutz, W., Winston, K.R., and Maleki, N.**, A system for stereotactic radiosurgery with a linear accelerator, *Int. J. Radiat. Oncol. Biol. Phys.*, 14, 373, 1988.

113.  **Lutz, W.R.**, The University of Arizona, personal communication, 1991.

114.  **Lyman, J.T., Phillips, M.H., Frankel, K.A., and Fabrikant, J.I.**, Stereotactic frame for neuroradiology and charged particle Bragg peak radiosurgery of intracranial disorders, *Int. J. Radiat. Oncol. Biol. Phys.*, 16, 1615, 1989.

115.  **Maitz, A.H., Lunsford, L.D., Wu, A., Lindner, G., and Flickinger, J.C.**, Shielding requirements on-site loading and acceptance testing of the Leksell gamma knife, *Int. J. Radiat. Oncol. Biol. Phys.*, 18, 469, 1990.

116. **Marin-Grez, M.**, High dose percutaneous stereotactic irradiation of solitary brain metastases using a 15 MeV linear accelerator, *Int. J. Radiat. Oncol. Biol. Phys.*, 15 (Suppl. 1), 231, 1988.

117. **Nedzi, L.A., Kooy, H., Alexander E., Gelman, R.S., Loeffler, J.S.**, Variables associated with the development of complications from radiosurgery of intracranial tumors, *Int. J. Radiat. Oncol. Biol. Phys.*, 21, 591, 1991.

118. **Neeranjun, W.S., Pennington, E.C.**, The University of Iowa, personal communication, 1992.

119. **Peters, T.M., Clark J., Pike, B., Drangova, M., and Olivier A.**, Stereotactic surgical planning with magnetic resonance imaging, digital subtraction angiography and computed tomography, *Appl. Neurophysiol.*, 50, 33, 1987.

120. **Phillips, M.H., Frankel, K.A., Lyman, J.T., Fabrikant, J.I., and Levy, R.P.**, Comparison of different radiation types and irradiation geometries in stereotactic radiosurgery, *Int. J. Radiat. Oncol. Biol. Phys.*, 18, 211, 1990.

121. **Phillips, M.H., Frankel, K.A., Lyman, J.T., Fabrikant, J.I., and Levy, R.P.**, Heavy charged-particle stereotactic radiosurgery: cerebral angiography and CT in the treatment of intracranial vascular malformations, *Int. J. Radiat. Oncol. Biol. Phys.*, 17, 419, 1989.

122. **Pike, B., Peters, T.M., Podgorsak, E.B., Pla, C., Olivier, A., and de Lotbinière, A.**, Stereotactic external beam calculations for radiosurgical treatment of brain lesions, *Appl. Neurophysiol.*, 50, 269, 1987.

123. **Pike, B., Podgorsak, E.B., Peters, T.M., and Pla, C.**, Dose distributions in dynamic stereotactic radiosurgery, *Med. Phys.*, 14, 780, 1987.

124. **Pike, G.B., Podgorsak, E.B., Peters, T.M., Pla, C., Olivier, A., and Souhami, L.**, Dose distributions in radiosurgery, *Med. Phys.*, 17, 296, 1990.

125. **Podgorsak, E.B., Olivier, A., Pla, M., Hazel, J., de Lotbinière, A., and Pike, B.**, Physical aspects of dynamic stereotactic radiosurgery, *Appl. Neurophysiol.*, 50, 263, 1987.

126. **Podgorsak, E.B., Olivier, A., Pla, M., Lefebvre, P.Y., and Hazel, J.**, Dynamic stereotactic radiosurgery, *Int. J. Radiat. Oncol. Biol. Phys.*, 14, 115, 1988.

127. **Podgorsak, E.B., Pike, G.B., Olivier, A., Pla, M., and Souhami, L.**, Radiosurgery with high energy photon beams: a comparison among techniques, *Int. J. Radiat. Oncol. Biol. Phys.*, 16, 857, 1989.

128. **Podgorsak, E.B., Pike, G.B., Pla, M., Olivier, A., and Souhami, L.**, Radiosurgery with photon beams: physical aspects and adequacy of linear accelerators, *Radiotherap. and Oncol.*, 17, 349, 1990.

129. **Rice, R.K., Hansen, J.L., Svensson, G.K., and Siddon, R.L.**, Measurements of dose distributions in small beams of 6 MV x-rays, *Phys. Med. Biol.* 32, 1087, 1987.

130. **Saunders, W.M., Winston, K.R., Siddon, R.L., Svensson, G.H., Kijewski, P.K., Rice, R.K., Hansen, J.L., and Barth N.H.**, Radiosurgery for arteriovenous malformations of the brain using a standard linear accelerator: rationale and technique, *Int. J. Radiat. Oncol. Biol. Phys.*, 15, 441, 1988.

131.  **Schell, M.C., Smith V., Larson, D.A., Wu, A., and Flickinger, J.C.**, Evaluation of radiosurgery techniques with cumulative dose volume histograms in linac-based stereotactic external beam irradiation, *Int. J. Radiat. Oncol. Biol. Phys.,* 20, 1325, 1991.

132.  **Schlegel, W., Pastyr, O., Bortfeld, T., Becker, G., Schad, L., Gademann, G., and Lorenz, W.J.**, Computer systems and mechanical tools for stereotactically guided conformation therapy with linear accelerators, *Int. J. Radiat. Oncol. Biol. Phys.,* 24, 781, 1992.

133.  **Serago, C.F., Houdek, P.V., Hartmann, G.H., Saini, D.S., Serago, M.E., and Kaydee, A.**, Tissue maximum ratios (and other parameters) of small circular 4, 6, 10, 15 and 24 MV x-ray beams for radiosurgery, *Phys. Med. Biol.,* 37, 1943, 1992.

134.  **Serago, C.F., Lewin, A.A., Houdek, P.V., Gonzales-Arias, S., Abitbol, A.A., Marcial-Vega, V.A., Pisciotti, V., and Schwade, J.G.**, Improved linac dose distributions for radiosurgery with elliptically shaped fields, *Int. J. Radiat. Oncol. Biol. Phys.,* 21, 1321, 1991.

135.  **Siddon, R.L. and Barth, N.H.**, Stereotaxic localization of intracranial targets, *Int. J. Radiat. Oncol. Biol. Phys.,* 13, 1241, 1987.

136.  **Souhami, L., Olivier, A., Podgorsak, E.B., Hazel, J., Pla, M., and Tampieri, D.**, Dynamic stereotactic radiosurgery in arteriovenous malformation, *Cancer,* 66, 15, 1990.

137.  **Souhami, L., Olivier, A., Podgorsak, E.B., Pla, M., and Pike, G.B.**, Radiosurgery of cerebral arteriovenous malformations with the dynamic stereotactic irradiation, *Int. J. Radiat. Oncol. Biol. Phys.,* 19, 775, 1990.

138.  **Spaulding, C.A. and Berk, H.W.**, Stereotaxic radiosurgery in the treatment of arteriovenous malformations, *Applied Radiology,* 11, July, 1989.

139.  **Steiner, L.**, Treatment of arteriovenous malformations by radiosurgery, in *Intracranial Arteriovenous Malformations,* Wilson and Stein, Eds., Williams & Wilkins, Baltimore, 1984.

140.  **Sturm, V., et al.**, Stereotactic percutaneous single dose irradiation of brain metastases with a linear accelerator, *Int. J. Radiat. Oncol. Biol. Phys.,* 13, 279, 1987.

141.  **Tsai, J.-S., Buck, B.A., Svensson, G.K., Alexander III, E., Cheng, C.-W., Mannarino, E.G., and Loeffler, J.S.**, Quality assurance in stereotactic radiosurgery using a standard linear accelerator, *Int. J. Radiat. Oncol. Biol. Phys.,* 21, 737, 1991.

142.  **Walton, L., Bomford, C.K., and Ramsden, D.**, The Sheffield stereotactic radiosurgery unit: physical characteristics and principles of operation, *Br. J. Radiol.,* 60, 897, 1987.

## D.  INTRAOPERATIVE RADIOTHERAPY

143.  **Bagne, F.R., Dobelbower Jr., R.R., Milligan A.J., and Bronn, D.G.**, Treatment of cancer of the pancreas by intraoperative electron beam therapy: physical and biological aspects, *Int. J. Radiat. Oncol. Biol. Phys.,* 16, 231, 1989.

144. **Bagne, F.R., Samsami, N., and Dobelbower Jr., R.R.**, Radiation contamination and leakage assessment of intraoperative electron applicators, *Med. Phys.*, 15, 530, 1988.

145. **Biggs, P.J. and Wang, C.C.**, An intra-oral cone for an 18 MeV linear accelerator, *Int. J. Radiat. Oncol. Biol. Phys.*, 8, 125, 1982.

146. **Biggs, P.J., Epp, E.R., Ling, C.C., Novack, D.H., and Michaels, H.B.**, Dosimetry, field shaping and other considerations for intra-operative electron therapy, *Int. J. Radiat. Oncol. Biol. Phys.*, 7, 875, 1981.

147. **Calvo, F.A., Azinovic, I., and Escudé, L.**, Intraoperative radiotherapy in cancer management, *Appl. Radiol.*, 15, January, 1992.

148. *Current Problems in Cancer*, Hickey, R.C., Ed., Year Book Medical Publishers, Inc., Chicago, IL, 1983.

149. **Dahl, R.A. and McCullough E.C.**, Determination of accurate dosimetric parameters for beveled intraoperative electron beam applicators, *Med. Phys.*, 16, 130, 1989.

150. **Fraass, B.A., Miller, R.W., Kinsella, T.J., Sindela, W.F. Harrington, F.S., Yerkel, K., Van de Geijn, J., and Glatstein, D.**, Intra operative radiation therapy at the National Cancer Institute: technical innovations and dosimetry, *Int. J. Radiat. Oncol. Biol. Phys.*, 11, 1299, 1985.

151. **Gillette, E.L., Powers, B.E., McChesney, S.L., Park, R.D., and Withrow, S.J.**, Response of aorta and branch arteries to experimental intraoperative irradiation, *Int. J. Radiat. Oncol. Biol. Phys.*, 17, 1247, 1989.

152. *Intraoperative Radiation Therapy*, Abe, M. and Takahashi, M., Eds., Pergamon Press, New York, 1990.

153. *Intraoperative Radiation Therapy*, Dobelbower, R.R., Abe, M., Eds., CRC Press, Inc., Boca Raton, FL, 1989.

154. **Jani, S.K.**, The University of Iowa Hospitals and Clinics, Iowa City, IA, personal communication, 1987.

155. **Jones, D.**, Apparatus, technique, and dosimetry of intraoperative electron beam therapy, *Front Radiat. Ther. Oncol.*, 25, 233, 1991.

156. **Jones, D., Taylor, E., Travaglini, J., and Vermeulen, S.**, A non-contacting intraoperative electron cone apparatus, *Int. J. Radiat. Oncol. Biol. Phys.*, 16, 1643, 1989.

157. **McCullough, E.C. and Anderson J.A.**, The dosimetric properties of an application system for intraoperative electron-beam therapy utilizing a Clinac-18 accelerator, *Med. Phys.*, 9, 261, 1982.

158. **McCullough, E.C. and Gunderson, L.L.**, Energy as well as applicator size and shaped utilized in over 200 intraoperative electron beam procedures, *Int. J. Radiat. Oncol. Biol. Phys.*, 15, 1041, 1988.

159. **McCullough, E.C., Biggs, P.J.**, Intraoperative electron beam radiation therapy, AAPM, Monograph 15, American Institute of Physics, 1986, 333.

160. **Nelson, C.E., Cook, R., and Rakfal, S.**, The dosimetric properties of an intraoperative radiation therapy applicator system for a Mevatron-80, *Med. Phys.*, 16, 794, 1989.

161. **Nyerick, G.E., Ochran, T.G., Boyer, A.L., and Hogstrom, K.R.**, Dosimetry characteristics of metallic cones for intraoperative radiotherapy, *Int. J. Radiat. Oncol. Biol. Phys.*, 21, 501, 1991.

162.  **Palta, J.R. and Suntharalingam, N.**, A non-docking intraoperative electron beam application system, *Int. J. Radiat. Oncol. Biol. Phys.*, 17, 411, 1989.

163.  **Shaw, E.G., Blackwell, R., McCullough, E.C., and Gunderson, L.L.**, Resident essay award: matching intraoperative electron-beam fields: dosimetric and clinical considerations, *Int. J. Radiat. Oncol. Biol. Phys.*, 13, 1303, 1987.

164.  **Tepper, J.E., Gunderson, L.L., Goldosn, A.L., Kinsella, T.J., Shipley, W.U., Sindelar, W.F., Wood, W.C., and Martin, J.K.**, Quality control parameters of intraoperative radiation therapy, *Int. J. Radiat. Oncol. Biol. Phys.*, 12, 1687, 1986.

165.  **Wilson, D.L., Sharma, S.C., and Jose, B.**, An intracavitary cone system for electron beam therapy using a Therac 20 linear accelerator, *Int. J. Radiat. Oncol. Biol. Phys.*, 12, 1107, 1986.

Chapter 4

# DATA ON BRACHYTHERAPY DOSIMETRY

In recent years, several significant changes have occurred in brachytherapy dosimetry. New radioisotopes exhibiting longer half life and emitting low energy photons have been studied and employed. Methods to calibrate brachytherapy sources are being standardized. The ultimate goal has been to determine, with reasonable confidence, the dose rates around sources implanted in tissue. Two recently addressed issues are:

1. What is the most suitable method to specify the strength of a commercially available radioactive source?
2. What is the dose distribution around a source of known strength?

Historically, the source strength has been specified in mg of radium, mg Ra Eq., mCi, or mCi-apparent. Recently, the AAPM Task Group has recommended that the strength of a source be specified in terms of exposure rate in free space or air kerma rate at certain reference point around the source.[12] This reference point may be at one meter away along transverse axis of the source. The user would then convert this exposure or air kerma rate to absorbed dose in tissue at certain practical point, for example, at 1 cm from the source along its transverse axis.

The dose distribution around a source needs to be carefully measured. It depends upon:

1. inverse-square law (i.e., distance)
3. geometry of sealed source, including its wall material and thickness
3. absorption in tissue, and
4. scattering by tissue

Geometric factors such as finite source size and its design may cause the dose to be direction-dependent (i.e., anisotropic). For small seed sources, this effect is small and may be described by a factor called "anisotropy factor". For linear sources, such as Cs-137 tubes, the anisotropy in dose is quite pronounced, and, therefore, the dose around them is tabulated as a function of distance and direction. The combined effects of absorption and scatter within the medium (tissue) can be described by a parameter called radial dose function, $g(r)$ which is defined later in this section. An excellent review of the dosimetric approaches in brachytherapy is given by Ravi Nath in a recently published AAPM Monograph.[2]

In this chapter, we describe the dosimetric characteristics of 13 radioisotopes that are either in use or exhibit a potential for use in brachytherapy. Important dosimetric parameters are defined as follows.

## 1.  Exposure rate constant

This is defined as the exposure rate in R/hr in *air* at a distance of 1 cm from a point source of 1 mCi in activity. This definition includes radiation exposure resulting from gamma rays, as well as other photons, such as characteristic X rays and internal bremsstrahlung.

This parameter, although old, has been useful in its practical applications. ICRU-33[7] recommends that it be replaced by air kerma rate constant as defined below. Unit of exposure rate constant is R cm$^2$ mCi$^{-1}$ hr$^{-1}$.

## 2.  Air kerma rate constant

This is the *air* kerma rate in cGy/hr at 1 m from a point source having an activity of 1 mCi. Its practical unit is cGy cm$^{-2}$ mCi$^{-1}$ hr$^{-1}$.

Air kerma rate constant = 0.876 cGy/R. Exposure rate constant.

## 3.  Dose rate constant

This is the absorbed dose rate in tissue in cGy/hr per unit source activity (mCi) or unit air kerma strength (V→U) at 1 cm along the transverse axis of the seed and has units of cGy mCi$^{-1}$ hr$^{-1}$ or cGy U$^{-1}$ hr$^{-1}$.

The dose rate constant applies to a real physical source rather than a point source of a particular radionuclide. Any change in source structure (e.g., material of encapsulation or its thickness) may result in a different dose rate in tissue at one cm away for the same air kerma strength, and hence, a different value of dose rate constant.

## 4.  f-factor

This converts exposure in air to absorbed dose in the medium, and has units of cGy/R.

## 5.  Radial dose function *g(r)*

This is the ratio of dose at distance *r* to dose at 1 cm in tissues with the effect of the inverse square law removed; both the points being on the transverse axis of the source. The *g(r)* accounts for absorption and scatter along the transverse axis and, by definition, is unity at 1 cm. It is experimentally obtained from the depth dose data of a sealed source.

For each radionuclide, we have attempted to include the latest available dosimetric data. For further details on sealed sources and parameters, the reader is referred to the bibliography at the end of this chapter.

## LIST OF FIGURES

## LIST OF TABLES

## A.  DATA ON Am-241

1.  *Name:* Americium (atomic number: 95; atomic mass: 241)

2.  **Production of Radionuclide**

    Am-241 is a radioactive metal of the Actinide series, and is a transuranic element. All isotopes of americium are radioactive, and must be produced synthetically. Am-241 is produced by bombardment of Pu-241 (plutonium, $Z = 94$) with neutrons in a nuclear reactor.

3.  **Decay Modes**

    Am-241 decays to Np-237 (neptunium, $Z = 93$) by emission of alpha particles. This decay also leads to emission of one Auger electron, 14 Coster-Kronig electrons (4–93 keV), and 139 known photons. The most prominent photons are 14 keV (43%) and 60 keV (36%). The alphas, electrons, and low energy photons are shielded by encapsulation of the source. Therefore, the primary radiation from a sealed Am-241 is 60 keV photons.

4.  *Average Photon Energy:* 60 keV

5.  *Half Life:* 432.2 years

6.  *Half Value Layer:* 0.12 mm lead

7.  *Available Forms for Brachytherapy*

    At present, the Am-241 sealed sources are not available commercially. A few prototype sources in the forms of flat disc and cylindrical tube were manufactured for dosimetric analysis.

8.  *Available Source Strengths*

    The prototype sources' activity was in the range of ~0.1 Ci to ~10 Ci.

9.  *Dosimetric Parameters*

    Exposure rate constant:  0.122 R cm$^2$ mCi$^{-1}$ hr$^{-1}$
    Air kerma rate constant: 0.107 cGy cm$^2$ mCi$^{-1}$ hr$^{-1}$

    The radial dose function for Am-241 seed source is given in Table 4.1.

10. *References:* See References 14 through 20 at the end of this chapter.

## B. DATA ON Au-198

*1.* *Name:* Gold (atomic number: 79; atomic mass: 198)

*2.* *Production of Radionuclide*

Au-198 is produced in a nuclear reactor when stable Au-197 absorbs a neutron.

*3.* *Decay Modes*

Au-198 decays to excited states of Hg-198 (mercury, $Z = 80$) via negative beta emission. The Hg-198 nucleus decays to its ground state by gamma emissions. This gamma radiation from sealed Au-198 sources is utilized for radiotherapy.

| Maximum beta energies: | 0.29 MeV (1.2%) |
|---|---|
| | 0.96 MeV (98.8%) |
| | 1.37 MeV (0.02%) |
| Gamma energies: | 0.412 MeV (94.7%) |
| | 0.676 MeV (1%) |
| | 1.09 MeV (0.2%) |

Note:  4% of gammas undergo internal conversion which is followed by Hg fluorescent X rays.

*4.* *Average Photon Energy*: 0.416 MeV

*5.* *Half Life*: 2.697 days

*6.* *Half Value Layer*: ~3 mm lead

*7.* *Available Forms for Brachytherapy*

In the past, Au-198 was available in colloidal solution which utilized beta emission for radiotherapy. At present, this isotope is commercially available in the U.S. primarily in the form of small seeds.

*8.* *Available Source Strengths*: ≤25 mCi

## 9.  *Schematic Drawing of Sealed Sources*

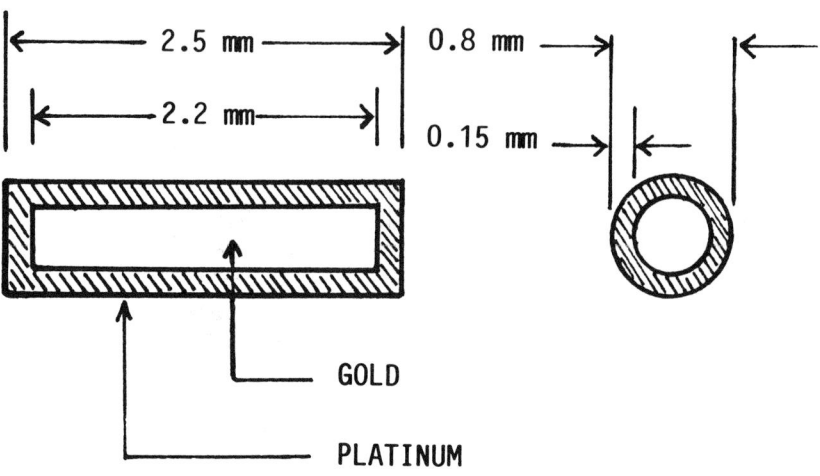

**FIGURE 4.1.**  Schematic drawing of Au-198 seed. (Best Industries, Springfield, VA.)

## 10.  *Dosimetric Parameters*

Exposure rate constant:    2.34  R cm$^2$ mCi$^{-1}$ hr$^{-1}$
Air kerma rate constant:  2.05  cGy cm$^2$ mCi$^{-1}$ hr$^{-1}$
Dose rate constant:          2.31  cGy mCi$^{-1}$ hr$^{-1}$

The radial dose function for Au-198 seed source is given in Table 4.1.

## 11.  *References:* See References 21 through 28 at the end of this chapter.

## C. DATA ON Cf-252

1. *Name:* Californium (atomic number: 98; atomic mass: 252)

2. *Production of Radionuclide*

   Cf-252 is produced by irradiating Pu-239 (plutonium, $Z = 94$) and higher transuranium elements in a nuclear reactor operating at a very high neutron flux. By a sequence of neutron capture and beta decay reactions, Pu-239 is converted in turn to americium (Am), curium (Cm), berkelium (Bk) and, finally, californium (Cf). Overall process yield is less than 1% due to fission losses. At the end of irradiation, the Cf-252 is separated chemically from the remaining intermediates and fission products.

3. *Decay Modes*

   Cf-252 decays by spontaneous fission (3.1%) and alpha emission (96.9%). Although, most of Cf-252 nuclei decay via alpha emission, the fission emits an average of 3.76 neutrons per fission event. The fission products decay by beta and gamma emission. The neutrons and gammas are used for radiotherapy, with special emphasis on neutrons.

4. *Average Neutron Energy*: 2.35 MeV

   Modal neutron energy: 1.5 MeV
   Average photon energy: 1 MeV

5. *Half Life*: 2.65 years

6. *Shielding requirement*

   The neutron shielding is of primary concern, and is achieved by use of hydrogenous materials to thermalize neutrons. The intensity of capture gamma rays (2.24 MeV) can be reduced by adding boron to material for neutron shielding, which produces (n,γ) reaction that competes with gamma production. Lead shielding is needed to reduce photon exposure levels around the sources.

7. *Available Forms for Brachytherapy*

   The U.S. Atomic Energy Commission manufactured sealed sources of Cf-252 in 1965 as part of a feasibility study on therapeutic use of Cf-252. The sources produced were in the form of seeds, needles, and tubes. At present, Cf-252 sources are not available commercially.

8. *Available Source Strengths*

   The prototype sources had activity in the range of 50–150 micrograms.

## 9. Schematic Drawing of Sealed Sources

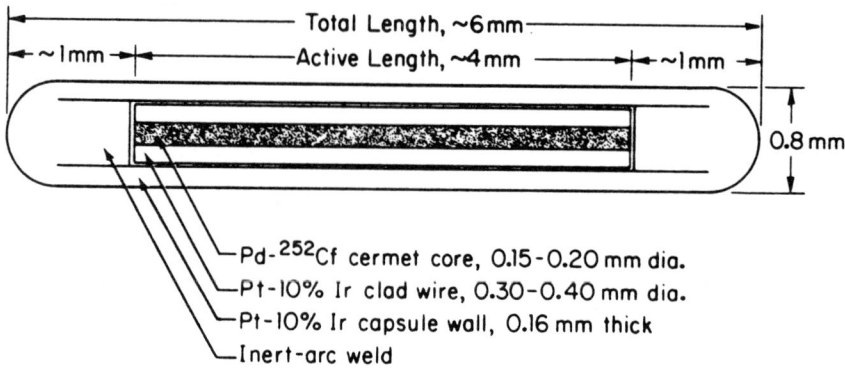

**FIGURE 4.2.** Schematic drawing of Cf-252 seed source. (Reprinted from Permar, D.H., Cf-252 neutron sources for interstitial afterloading, *Int. J. Radiat. Oncol. Biol. Phys.*, 1, 1003, 1976. With permission.)

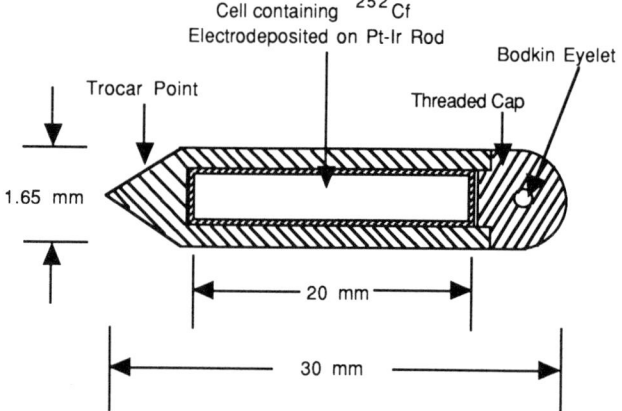

**FIGURE 4.3.** Schematic drawing of Cf-252 needle source.

## 10. Dosimetric Parameters

Air kerma rate constants:  neutrons: 1.99   cGy cm$^2$ mg$^{-1}$ hr$^{-1}$
gammas: 1.08   cGy cm$^2$ mg$^{-1}$ hr$^{-1}$

The neutron/gamma dose ratio varies from point-to-point around a source, ranging from 1.3–2.6.

The dose distribution around Cf-252 linear sources is given in Table 4.2.

## 11. References: See References 29 through 45 at the end of this chapter.

## D. DATA ON Cs-137

*1.* **Name**: Cesium (atomic number: 55; atomic mass: 137)

*2.* **Production of Radionuclide**

Cs-137 is a product of nuclear fission and is produced in quantity as a byproduct of nuclear fuel reprocessing. It is a toxic alkali metal of low melting point, but is available as stable compounds such as chloride or sulfate.

*3.* **Decay Modes**

Cs-137 decays via negative beta emission to either a metastable state of Ba-137 (barium, $Z = 56$) or stable Ba-137. For 93.5% of decays, the result is metastable state of Ba-137, which eventually reaches its ground state by emitting a 0.662-MeV gamma ray. This gamma radiation from sealed Cs-137 sources is used for radiotherapy.

*4.* **Average Photon Energy**: 0.662 MeV

*5.* **Half Life**: 30.0 years

*6.* **Half Value Layer**: 6 mm lead

*7.* **Available Forms for Brachytherapy**

Sealed sources in the form of seeds, needles, and tubes.

*8.* **Available Source Strengths**: $\leq$100 mCi

*9.* **Schematic Drawing of Sealed Sources**

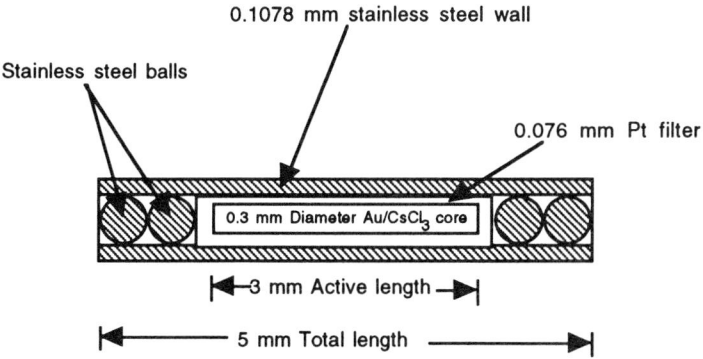

**FIGURE 4.4.** Schematic drawing of Cs-137 seed source.
(Radiation Therapy Resources, Valencia, CA.)

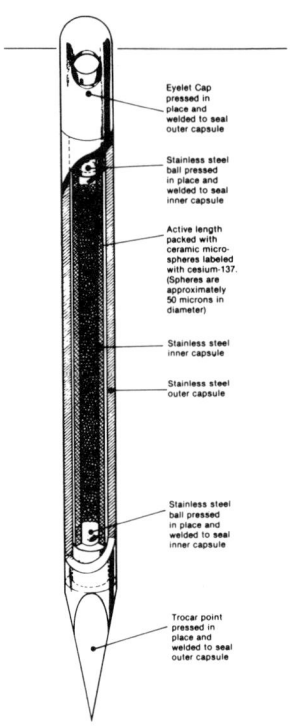

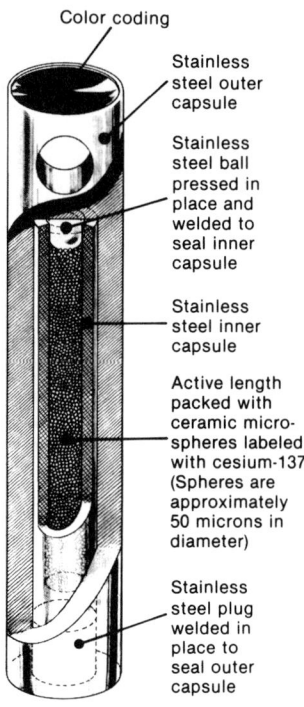

**FIGURE 4.5.** Schematic drawing of Cs-137 needle source. (3M Co., St. Paul, MN.)

**FIGURE 4.6.** Schematic drawing of Cs-137 tube source. (3M Co., St. Paul, MN.)

## 10. Dosimetric Parameters

Exposure rate constant:    3.28  R cm$^2$ mCi$^{-1}$ hr$^{-1}$
Air kerma rate constant:   2.89  cGy cm$^2$ mCi$^{-1}$ hr$^{-1}$
Dose rate constant
   Point source:        3.09  cGy mCi$^{-1}$ hr$^{-1}$
   Seed source:         3.17  cGy mCi$^{-1}$ hr$^{-1}$

The radial dose function for Cs-137 seed source is given in Table 4.1.

Dose distribution around Cs-137 linear source is given in Table 4.3.

Dose rates at reference points for a gynecological application using Cs-137 are given in Table 4.4.

## 11. References: See References 46 through 61 at the end of this chapter.

## E. DATA ON I-125

1. *Name:* Iodine (atomic number: 53; atomic mass: 125)

2. *Production of Radionuclide*

   I-125 is produced in a nuclear reactor when Xe-124 (xenon, Z = 54) absorbs a neutron to become Xe-125, which then undergoes electron capture.

3. *Decay Modes*

   I-125 nuclei decay via electron capture to excited state of Te-125 (tellurium, Z = 52). The Te-125 reaches its ground (stable) state by either internal conversion (93%) or gamma emission (7%). Both electron capture as well as internal conversion remove atomic electrons and lead to the production of characteristic X rays, mainly in the energy range of 27.4–35 keV.

4. *Average Photon Energy:* 28 keV

5. *Half Life:* 59.6 days

6. *Half Value Layer:* 0.025 mm lead

7. *Available Forms for Brachytherapy*

   Sealed sources of I-125 isotope are available in the form of seeds.

8. *Available Source Strengths:* ≤40 mCi

9. *Schematic Drawing of Sealed Sources*

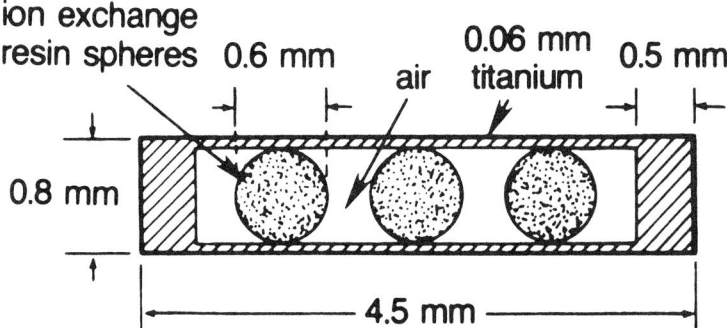

**FIGURE 4.7.** Schematic drawing of I-125 seed source. (Model 6702, 3M Co., St. Paul, MN.)

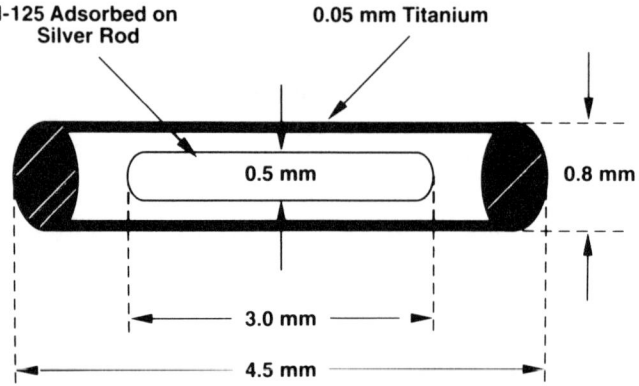

**FIGURE 4.8.** Schematic drawing of I-125 seed source. (Model 6711, 3M Co., St. Paul, MN.)

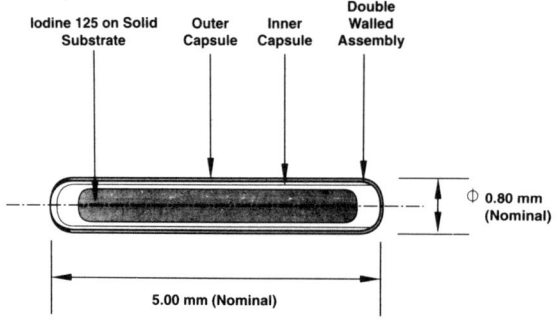

**FIGURE 4.9.** Schematic drawing of I-125 seed source.
(Model 2300, Best Industries, Springfield, VA.)

## 10. *Dosimetric Parameters*

| | |
|---|---|
| Exposure rate constant: | 1.45  R cm$^2$ mCi$^{-1}$ hr$^{-1}$ |
| Air kerma rate constant: | 1.27  cGy cm$^2$ mCi$^{-1}$ hr$^{-1}$ |
| Dose rate constant: | 1.16  cGy mCi$^{-1}$ hr$^{-1}$ (3M-Model 6702) |
| | 1.08  cGy mCi$^{-1}$ hr$^{-1}$ (3M-Model 6711) |

| | |
|---|---|
| Dose anisotropy factor: | 0.96  (3M-Model 6702 seeds) |
| | 0.94  (3M-Model 6711 seeds) |
| Photon fluence anisotropy factor: | 0.87  (3M-Model 6711 seeds) |
| | 0.92  (Best -Model 2300 seeds) |

The radial dose function for I-125 seed sources is given in Table 4.1.

## 11. *References:* See References 62 through 102 at the end of this chapter.

## F. DATA ON I-131

1. *Name:* Iodine (atomic number: 53; atomic mass: 131)

2. *Production of Radionuclide*

   I-131 is produced in a nuclear reactor when Te-130 (tellurium, $Z = 52$) absorbs a neutron to become Te-131, which then undergoes negative beta decay to become I-131. Alternately, I-131 can be obtained as a reactor byproduct from fissioning U-235 (uranium, $Z = 92$).

3. *Decay Modes*

   I-131 decays to excited states of Xe-131 (xenon, $Z = 54$) via negative beta emission with maximum energy of 0.25 MeV (2.8%), 0.335 MeV (9.3%), 0.608 MeV (87.2%) and 0.812 MeV (0.7%). Xe-131 reaches the stable (ground) state by gamma emission in the energy range of 80–724 keV; the principal energy being 364 keV.

4. *Principal Gamma Energy:* 364 keV
   Mean beta energy: 192 keV

5. *Half Life:* 8.05 days

6. *Half Value Thickness:* 6.3 cm in water

7. *Available Forms for Radiotherapy*

   I-131 is utilized as a therapeutic radiopharmaceutical in the form of sodium iodide for the treatment of hyperthyroidism and thyroid cancer (including metastases). Sodium iodide (I-131) is available in either capsular or liquid form.

   Other parameters:
   Typical dosage for hyperthyroidism is 5–10 mCi of I-131 sodium iodide.
   Typical dosage for thyroid cancer is 100–200 mCi of I-131 sodium iodide.

8. *References:* See References 103 through 110 at the end of this chapter.

## G.  DATA ON Ir-192

1.  *Name:* Iridium (atomic number: 77; atomic mass: 192)

2.  **Production of Radionuclide**

    Ir-192 is formed in a nuclear reactor when stable Ir-191 (37% natural abundance) absorbs a neutron. Shipment of Ir-192 seeds to users is delayed until about two weeks after the seeds are removed from the reactor. This is to ensure that the amount of Ir-194 (half life of 18 hours) produced by activation of Ir-193 (63% natural abundance) becomes negligible.

3.  **Decay Modes**

    Ir-192 decays via negative beta emission (95.6%) to Pt-192 (platinum, $Z = 78$) and via electron capture (4.4%) to Os-192 (osmium, $Z = 76$). The excited nuclei of Pt-192 and Os-192 emit a number of gamma rays in the energy range of 0.136–1.06 MeV;  the primary emission  being in the 0.3–0.6 MeV range.

    Maximum beta energy:   0.67 MeV
    Gamma energies:            0.206 MeV (0.034/decay)
                                        0.296 MeV (0.291/decay)
                                        0.308 MeV (0.298/decay)
                                        0.317 MeV (0.831/decay)
                                        0.468 MeV (0.476/decay)
                                        0.485 MeV (0.032/decay)
                                        0.589 MeV (0.044/decay)
                                        0.604 MeV (0.081/decay)
                                        0.612 MeV (0.052/decay)

    Notes:  Average number of gammas per decay is 2.2.
              Pt and Os fluorescent X rays of mainly 0.067 MeV accompany the gamma emission.

4.  *Average Photon Energy:* 0.37 MeV

5.  *Half Life:* 74.2 days

6.  *Half Value Layer:* ~3 mm lead

7.  *Available Forms for Brachytherapy*

    Sealed Ir-192 sources are available in the form of seeds and wires.

8.  *Available Source Strengths:* ≤15 mCi

## 9. Schematic Drawing of Sealed Sources

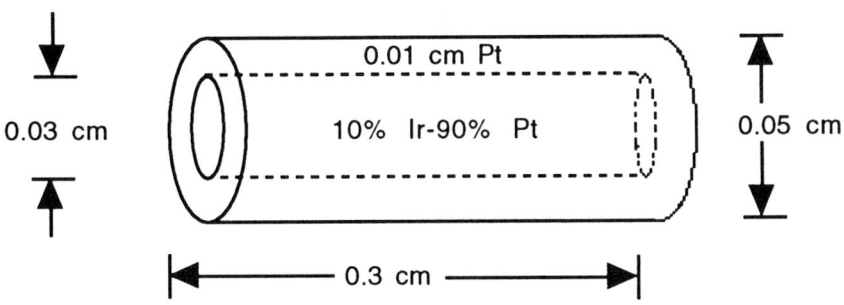

**FIGURE 4.10.** Schematic drawing of Ir-192 seed source.
(Alpha-Omega Services, Inc., Paramount, CA.)

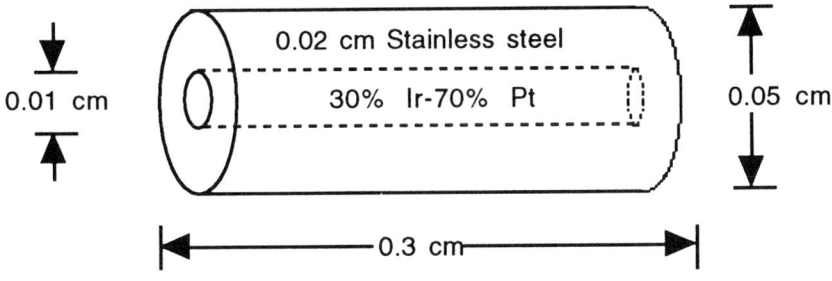

**FIGURE 4.11.** Schematic drawing of Ir-192 seed source.
(Best Industries, Inc., Springfield, VA.)

## 10. Dosimetric Parameters

Exposure rate constant:   4.69  R cm$^2$ mCi$^{-1}$ hr$^{-1}$
Air kerma rate constant:  4.11  cGy cm$^2$ mCi$^{-1}$ hr$^{-1}$
Dose rate constant:       4.55  cGy mCi$^{-1}$ hr$^{-1}$

The radial dose function for Ir-192 seed sources is given in Table 4.1.

## 11. References: See References 111 through 152 at the end of this chapter.

## H.  DATA ON P-32

1. *Name:* Phosphorus (atomic number: 15; atomic mass: 32)

2. **Production of Radionuclide**

   P-32 can be produced in a nuclear reactor by adding a neutron to P-31 nucleus (natural abundance = 100%).

3. **Decay Modes**

   P-32 decays by negative beta emission to S-32 (sulfur, $Z = 16$). The maximum beta energy is 1.71 MeV. P-32 is a pure beta emitter.

4. *Average Beta Energy:* 0.695 MeV

5. *Half Life:* 14.3 days

6. **Half Value Thickness in Tissue:**          0.8 mm
   Maximum penetration in tissue:     8 mm
   Maximum range in air:                     18 feet

7. *Available Forms for Radiotherapy*

   P-32 is utilized as therapeutic radiopharmaceutical in the form of sodium phosphate or chromic phosphate. The sodium phosphate P-32 is used intravenously for polycythemia vera and for relief of pain from bone metastases. The chromic phosphate P-32 is used in intracavitary applications for treating malignant effusion, cystic brain tumors and ovarian cancers.

8. *Dosimetric Considerations*

   • For treating malignant peritoneal effusion, the P-32 chromic phosphate is administered intraperitoneally. Typical dosage range is 10–20 mCi.

   • For treating malignant pleural effusion, the P-32 chromic phosphate is administered intrapleurally. Typical dosage range is 5–15 mCi.

   • For treating malignant pericardial effusion, the P-32 chromic phosphate is administered intrapericardially. Typical dosage range is 5–10 mCi.

   • For the treatment of cystic brain lesions, the P-32 chromic phosphate can be injected into the cyst with barbotage.[157] Typical dosage is 0.1–3 mCi to deliver 200 Gy to the cyst wall.

9. *References:* See References 153 through 159 at the end of this chapter.

## I.  DATA ON Pd-103

1.  *Name:* Palladium (atomic number: 46; atomic mass: 103)

2.  **Production of Radionuclide**

    Pd-103 is produced in a nuclear reactor when Pd-102 absorbs a neutron.

3.  **Decay Modes**

    Pd-103 decays via electron capture, largely to the first (90%) and second (10%) excited states of Rh-103 (rhodium, $Z = 45$) which becomes stable by the process of internal conversion. The first excited state of Rh-103 is an isomeric state with a 56 min half life. The electron capture as well as internal conversion give rise to characteristic X rays mainly in the energy range of 20–23 keV.

    Photon energies:   20.1 keV (K $\alpha$ X rays, 0.656/decay)
                       23.0 keV (K $\beta$ rays, 0.125/decay)
                       39.7 keV ($\gamma$ ray, 0.001/decay)
                       357 keV ($\gamma$ ray, 0.001/decay)

    Note: Average number of photons per decay is 0.8.

4.  *Average Photon Energy:* 21 keV

5.  *Half Life:* 17 days

6.  *Half Value Layer:* ~0.01 mm lead

7.  *Available Forms for Brachytherapy*

    Sealed sources of Pd-103 are available commercially in the form of seeds. Stable Pd-102 is first irradiated in a nuclear reactor to produce Pd-103. The Pd-103 purified by chemical separation is then electroplated on to the surface of two graphite cylinders which, in turn, are loaded into a titanium tube. Both ends of the tube are then sealed with titanium cups using a laser welding technique.

8.  *Available Source Strengths:* ≤5 mCi

### 9. Schematic Drawing of Sealed Sources

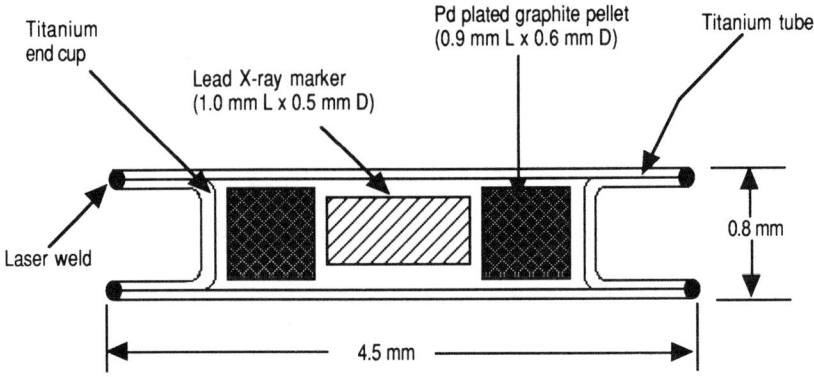

**FIGURE 4.12.** Schematic drawing of Pd-103 seed source.
(Model 200 Theragenics, Corp., Atlanta, GA.)

### 10. Dosimetric Parameters

Exposure rate constant:   1.48  R cm$^2$ mCi$^{-1}$ hr$^{-1}$
Air kerma rate constant:  1.30  cGy cm$^2$ mCi$^{-1}$ hr$^{-1}$
Dose rate constant:       0.95  cGy mCi$^{-1}$ hr$^{-1}$

The radial dose function for Pd-103 seed sources is given in Table 4.1.

### 11. References: See References 160 through 164 at the end of this chapter.

## J. DATA ON Ra-226

1. *Name:* Radium (atomic number: 88; atomic mass: 226)

2. *Production of Radionuclide*

   Ra-226 is a naturally occurring element. It is the sixth member of the uranium series, which starts with U-238 ($Z = 92$), and ends with stable Pb-206 (lead, $Z = 82$).

3. *Decay Modes*

   Ra-226 decays to Rn-222 (radon, $Z = 86$) with a half life of 1622 years. It, in turn, decays to Po-218 (polonium, $Z = 84$) with a half life of 3.83 days. Radioactive transformation continues for several more steps through alpha, beta and/or gamma emission. It terminates at stable Pb-206 (lead, $Z = 82$). At least 49 gamma rays are produced in the energy range of 0.184–2.45 MeV during this series of transformation. These gamma rays are utilized for radiotherapy.

4. *Average Photon Energy:* 0.83 MeV (with 0.5mm Pt filtration).

5. *Half Life:* 1622 years

6. *Half Value Layer:* 8 mm lead

7. *Available Forms for Brachytherapy*

   Ra-226, mostly in the form of radium sulfate or radium chloride salt, is mixed with an inert filler and loaded into gold cells about 1 cm long and 1 mm in diameter. The sealed cells are then loaded into the platinum sheath which, in turn, is sealed. The sealed Ra-226 sources were available in the form of needles and tubes. This isotope is now replaced by other isotopes such as Cs-137 for radiotherapy.

8. *Available Source Strengths:* ≤30 mg

9. *Dosimetric Parameters*

   | | | |
   |---|---|---|
   | Exposure rate constant: | 8.25 | R cm$^2$ mCi$^{-1}$ hr$^{-1}$ (0.5 mm Pt) |
   | Air kerma rate constant: | 7.31 | cGy cm$^2$ mCi$^{-1}$ hr$^{-1}$ (0.5 mm Pt) |
   | Air kerma rate constant: | 6.84 | cGy mCi$^{-1}$ hr$^{-1}$ (1.0 mm Pt) |
   | Activity/mass: | 0.988 | mCi/mg |

10. *References:* See References 165 through 174 at the end of this chapter.

## K.  DATA ON Sm-145

1.  *Name:* Samarium (atomic number: 62; atomic mass: 145)

2.  *Production of Radionuclide*

    It is produced by neutron irradiation of enriched Sm-144 in nuclear reactor.

3.  *Decay Modes*

    Sm-145 decays by electron capture to Pm-145(promethium; Z = 61), which, in turn, decays to stable Nd-145 (neodymium; Z = 60) by the same mechanism. X rays in the energy range of 38–61 keV are emitted from a sealed Sm-145 source.

    Photon energies:

| Sm-145 | Pm-145 |
|---|---|
| 38.2 keV (0.384 X rays/decay) | 36.9 keV (0.211 X rays/decay) |
| 38.7 keV (0.739 X rays/decay) | 37.4 keV (0.386 X rays/decay) |
| 43.8 keV (0.222 X rays/decay) | 42.2 keV (0.122 X rays/decay) |
| 44.9 keV (0.044 X rays/decay) | 43.3 keV (0.025 X rays/decay) |
| 61.4 keV (0.127 X rays/decay) | 67.2 keV (0.007 γ rays/decay) |
|  | 72.4 keV (0.022 γ rays/decay) |

    Note: Average number of photons per decay is 2.29.

4.  *Average Photon Energy:* 41 keV

5.  *Half Life:* 340 days

6.  *Half Value Layer:* 0.06 mm of lead

7.  *Available Forms for Brachytherapy*

    Presently, the sealed Sm-145 sources are not available commercially. A prototype source was produced with 2 mg of $Sm_2O_3$ encapsulated in titanium (4.5 mm ×0.8 mm; 0.05 mm wall thickness). See Ref. 175 for details.

8.  *Available Source Strengths*

    Activity of the prototype source was 3 mCi.

## 9. Illustration of Sealed Sources

**FIGURE 4.13.** Photograph of a Sm-145 prototype seed source. (Reprinted from Fairchild, R.G. et al., Samarium-145: a new brachytherapy source, *Phys. Med. Biol.,* 32, 847, 1987. With permission.)

## 10. Dosimetric Parameters

Exposure rate constant: 0.885   R cm$^2$ mCi$^{-1}$ hr$^{-1}$
Air kerma rate constant: 0.775   cGy cm$^2$ mCi$^{-1}$ hr$^{-1}$
Dose rate constant:      1.1   cGy mCi$^{-1}$ hr$^{-1}$

The radial dose function for Sm-145 seed source is given in Table 4.1.

## 11. References: See References 175 and 176 at the end of this chapter.

## L.  DATA ON Sr-90

1.  *Name:* Strontium (atomic number: 38; atomic mass: 90)

2.  **Production of Radionuclide**

    Sr-90 is produced as a result of fission reaction. This reaction is produced by bombarding certain high atomic number nuclei with neutrons. The nucleus, after absorbing the neutron, splits into two or more nuclei of medium atomic number. Sr-90 is a fission fragment produced in nuclear reactors.

3.  **Decay Modes**

    Sr-90 decays via negative beta emission to Y-90 (yttrium, $Z = 39$). The maximum beta energy is 0.546 MeV. Y-90 decays via same mechanism to stable Zr-90 (zirconium, $Z = 40$). The maximum beta energy from Y-90 is 2.283 MeV. Only a small amount of gamma radiation accompanies these decays. Ophthalmic applicators utilize Y-90 betas for radiotherapy.

4.  **Maximum Beta Energy:** 2.28 MeV

    Most frequent beta energy: 0.70 MeV

5.  **Half Life:** 28.5 years

    Half life of Y-90 is 64.2 hours. In a Sr-90 applicator, the Y-90 is continuously replenished through Sr-90 decay.

6.  **Half Value Thickness:** 0.125 cm in tissue

7.  **Available Forms for Brachytherapy**

    Sr-90 is employed in ophthalmic applicators which possess either flat or concave surface at the useful end of a cylindrical housing. Typical active area of a Sr-90 surface applicator is 6 to 10 mm in diameter.

8.  **Available Source Strengths**

    Typically 20 to 80 cGy/sec of surface dose rates.

9.  **Dosimetric Parameters**

    The percent depth dose from a 1 cm diameter flat applicator (Amersham-Model SIA.20) has been measured to be 100, 82, 63, 50, 36, 20, 10, 5 and 2 at corresponding depths of 0, 0.5, 1, 1.25, 2, 3, 4, 5 and 6 mm in water.[186]

10. **References:** See References 177 through 193 at the end of this chapter.

## M. DATA ON Yb-169

1. *Name:* Ytterbium (atomic number: 70; atomic mass: 169)

2. *Production of Radionuclide*

   Yb-169 can be produced in a nuclear reactor by neutron activation of ytterbium oxide ($Yb_2O_3$) enriched in Yb-168 isotope. Also produced during the activation is Yb-175, a contaminant radioisotope. It has a half life of 4.2 days and X ray emission in the range of 8–400 keV. The contamination level can be reduced by using highly enriched samples. In addition, the shipment of Yb-169 seeds can be delayed until Yb-175 is decayed to negligible levels.

3. *Decay Modes*

   Yb-169 decays by electron capture to stable Tm-169 (thulium, $Z = 69$), emitting X rays in the energy range of 50–308 keV. The principal X ray energies are 50, 51, 57, 63, 178, and 198 keV.

   Photon energies:

      49.8 keV (0.53 photons/decay)
      50.7 keV (0.94 photons/decay)
      57.5 keV (0.38 photons/decay)
      63.1 keV (0.48 photons/decay)
   109.8 keV (0.17 photons/decay)
   130.5 keV (0.11 photons/decay)
   177.2 keV (0.21 photons/decay)
   198.0 keV (0.35 photons/decay)
   307.7 keV (0.11 photons/decay)

4. *Average Photon Energy:* 93 keV

5. *Half Life:* 32 days

6. *Half Value Layer:* 0.2 mm of lead
                        4.0 cm of water

7. *Available Forms for Brachytherapy*

   At present, the sealed Yb-169 sources are not commercially available. Two prototype sealed sources in the form of seeds have been reportedly produced for dosimetric studies. (See Ref. 196 for details.)

8. *Available Source Strengths*

   The prototype seeds have had activity in the range of 1–10 mCi.

### 9. Schematic Drawing of Sealed Sources

0.838 mm

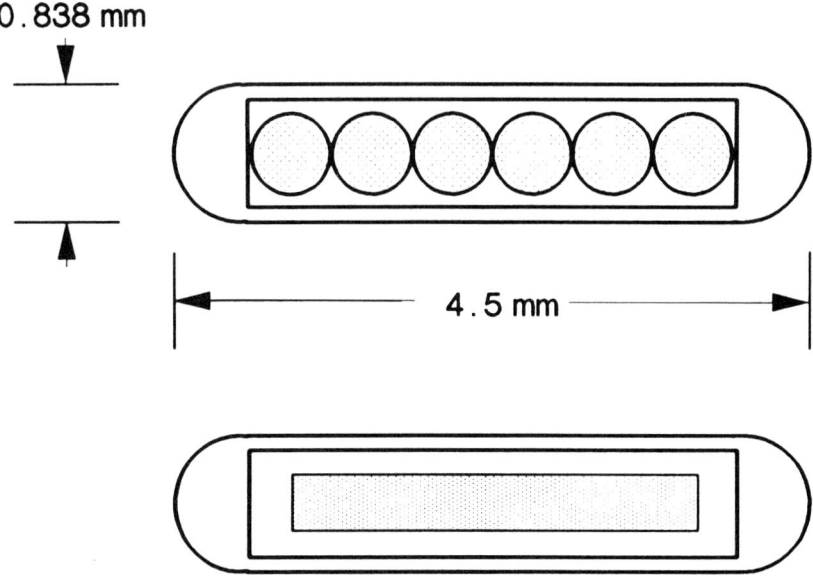

**FIGURE 4.14.** Schematic drawings of two types of Yb-169 seed sources. (Reprinted from Mason, D.L.D., Battista, J.J., Barnett, R.B., and Porter, A.T., Ytterbium-169: calculated physical properties of a new radiation source for brachytherapy, *Med. Phys.*, 19, 695, 1992. With permission.)

### 10. Dosimetric Parameters

| | | |
|---|---|---|
| Exposure rate constant: | 1.8 | R cm$^2$ mCi$^{-1}$ hr$^{-1}$ |
| Air kerma rate constant: | 1.58 | cGy cm$^2$ mCi$^{-1}$ hr$^{-1}$ |
| $F_{tissue}$: | 0.922 | cGy/R |
| $F_{bone}$: | 2.12 | cGy/R |

The radial dose function for Yb-169 seed sources is given in Table 4.1.

### 11. References: See References 194 through 198 at the end of this chapter.

## TABLE 4.1
### The Radial Dose Function for Small Brachytherapy Sources

| Radial Distance r (cm) | Radial Dose Function, g(r) | | | | | | | | | |
|---|---|---|---|---|---|---|---|---|---|---|
| | Am-241 | Au-198 | Cs-137 | I-125* | I-125 † | Ir-192 | Pd-103 | Sm-145 | Yb-169 | |
| 1.0 | 1.00 | 1.00 | 1.00 | 1.00 | 1.00 | 1.00 | 1.00 | 1.00 | 1.00 | |
| 2.0 | 1.01 | 1.01 | 0.99 | 0.86 | 0.84 | 1.03 | 0.54 | 1.02 | 1.10 | |
| 3.0 | 1.06 | 1.01 | 0.98 | 0.68 | 0.63 | 1.00 | 0.29 | 0.97 | 1.18 | |
| 4.0 | 1.10 | 1.01 | 0.97 | 0.53 | 0.47 | 0.97 | 0.16 | 0.92 | 1.21 | |
| 5.0 | 1.12 | 1.01 | 0.96 | 0.38 | 0.34 | 0.97 | 0.09 | 0.87 | 1.23 | |
| 6.0 | 1.11 | 1.01 | 0.95 | 0.29 | 0.26 | 0.97 | 0.05 | 0.78 | 1.23 | |
| 7.0 | 1.06 | 1.00 | 0.93 | 0.23 | 0.20 | 0.95 | 0.03 | 0.69 | 1.20 | |
| 8.0 | 0.99 | 0.98 | 0.91 | 0.20 | 0.16 | 0.94 | — | 0.61 | 1.17 | |
| 9.0 | 0.89 | 0.97 | 0.90 | — | — | 0.91 | — | 0.52 | 1.14 | |
| 10.0 | 0.79 | 0.95 | 0.88 | — | — | 0.87 | — | 0.44 | 1.10 | |

* 3M-Model-6702
† 3M-Model 6711

## TABLE 4.2
## Dose Rates in cGy/hr · μg Around a Cf-252 Source in Tissue
## (Active length = 1.5 cm; wall thickness = 0.25 mm Pt-Ir)

| Distance From Source Center Along its Length (cm) | Transverse Distance From Source Axis (cm) | | | | | | | | | | |
|---|---|---|---|---|---|---|---|---|---|---|---|
| | 0 | 0.5 | 1.0 | 1.5 | 2.0 | 2.5 | 3.0 | 3.5 | 4.0 | 4.5 | 5.0 |
| 0.0 | — | 8.15 | 2.64 | 1.25 | 0.71 | 0.46 | 0.32 | 0.23 | 0.17 | 0.14 | 0.11 |
| 0.5 | — | 6.84 | 2.33 | 1.16 | 0.68 | 0.44 | 0.31 | 0.22 | 0.17 | 0.13 | 0.11 |
| 1.0 | 5.85 | 3.37 | 1.63 | 0.93 | 0.59 | 0.40 | 0.28 | 0.21 | 0.16 | 0.13 | 0.10 |
| 1.5 | 1.39 | 1.45 | 1.01 | 0.68 | 0.47 | 0.34 | 0.25 | 0.19 | 0.15 | 0.12 | 0.10 |
| 2.0 | 0.65 | 0.76 | 0.63 | 0.49 | 0.37 | 0.28 | 0.22 | 0.17 | 0.14 | 0.11 | 0.09 |
| 2.5 | 0.38 | 0.46 | 0.42 | 0.35 | 0.29 | 0.23 | 0.19 | 0.15 | 0.12 | 0.10 | 0.09 |
| 3.0 | 0.25 | 0.31 | 0.29 | 0.26 | 0.22 | 0.19 | 0.16 | 0.13 | 0.11 | 0.09 | 0.08 |
| 3.5 | 0.18 | 0.22 | 0.21 | 0.20 | 0.17 | 0.15 | 0.13 | 0.11 | 0.10 | 0.08 | 0.07 |
| 4.0 | 0.13 | 0.16 | 0.16 | 0.15 | 0.14 | 0.12 | 0.11 | 0.10 | 0.08 | 0.07 | 0.06 |
| 4.5 | 0.10 | 0.12 | 0.13 | 0.12 | 0.11 | 0.10 | 0.09 | 0.08 | 0.07 | 0.06 | 0.06 |
| 5.0 | 0.08 | 0.10 | 0.10 | 0.10 | 0.09 | 0.08 | 0.08 | 0.07 | 0.06 | 0.06 | 0.05 |

From Robertson, J.S., Fairchild, R.G. and Atkins, H.L., Dosimetry of californium 252, *Radiol.*, 104, 393, 1972. With permission.

**TABLE 4.3**

**Dose Rates in cGy/hr/mg Ra. Eq. Cs-137 Source in Tissue**

**(Active length = 1.4 cm; wall thickness = 1.0 mm stainless steel)**

| Distance From Source Center Along its Length (cm) | Transverse Distance From Source Axis (cm) | | | | | | | | | |
|---|---|---|---|---|---|---|---|---|---|---|
| | 0.5 | 1.0 | 1.5 | 2.0 | 2.5 | 3.0 | 3.5 | 4.0 | 4.5 | 5.0 |
| 0.0 | 21.05 | 6.81 | 3.24 | 1.87 | 1.20 | 0.84 | 0.61 | 0.47 | 0.37 | 0.30 |
| 0.5 | 17.45 | 6.00 | 3.00 | 1.77 | 1.16 | 0.82 | 0.60 | 0.46 | 0.36 | 0.29 |
| 1.0 | 8.40 | 4.18 | 2.41 | 1.54 | 1.05 | 0.76 | 0.57 | 0.44 | 0.35 | 0.28 |
| 1.5 | 3.66 | 2.60 | 1.78 | 1.24 | 0.90 | 0.68 | 0.52 | 0.41 | 0.33 | 0.27 |
| 2.0 | 1.94 | 1.64 | 1.27 | 0.97 | 0.75 | 0.58 | 0.46 | 0.37 | 0.31 | 0.25 |
| 2.5 | 1.19 | 1.09 | 0.92 | 0.76 | 0.61 | 0.50 | 0.41 | 0.34 | 0.28 | 0.24 |
| 3.0 | 0.79 | 0.77 | 0.69 | 0.59 | 0.50 | 0.42 | 0.35 | 0.30 | 0.25 | 0.22 |
| 3.5 | 0.57 | 0.56 | 0.52 | 0.47 | 0.41 | 0.35 | 0.30 | 0.26 | 0.23 | 0.20 |
| 4.0 | 0.42 | 0.43 | 0.41 | 0.37 | 0.34 | 0.30 | 0.26 | 0.23 | 0.20 | 0.18 |
| 4.5 | 0.33 | 0.33 | 0.32 | 0.30 | 0.28 | 0.25 | 0.23 | 0.20 | 0.18 | 0.16 |
| 5.0 | 0.26 | 0.27 | 0.26 | 0.25 | 0.23 | 0.21 | 0.19 | 0.18 | 0.16 | 0.14 |

From Krishnaswamy, V., Dose distributions about [137]Cs sources in tissue, *Radiol.*, 105, 181, 1972. With permission.

# TABLE 4.4

## Dose Rates at Reference Points from an Intracavitary Gynecologic Application with Five Cs-137 Tube Sources

| Reference Point | Dose Rates in cGy/hr per mg·Ra·Eq·of Total Activity (study of 103 patients) | | |
| --- | --- | --- | --- |
| | Mean Value | Std. Deviation | Modal Value |
| Point A | 0.91 | 0.11 | 0.88 |
| Point B | 0.25 | 0.02 | 0.24 |
| Bladder | 0.61 | 0.20 | 0.50 |
| Rectum | 0.49 | 0.15 | 0.45 |
| Ext. illiac nodes | 0.17 | 0.02 | 0.17 |
| Common illiac nodes | 0.12 | 0.04 | 0.09 |
| Para-aortic nodes | 0.03 | 0.01 | 0.03 |
| Surface doses | | | |
| Mini ovoids | 12.4 | | |
| Small ovoids | 7.9 | | |
| Medium ovoids | 5.3 | | |
| Large ovoids | 3.8 | | |

Note: Doses on the ovoid surface are given in cGy/hour per mgRaEq Cs-137 activity loaded in that ovoid. Values apply to a typical five source cervical application. Ovoid attenuation=6%.
Mini=1.6 cm diameter; Small=2.0 cm diameter;
Medium=2.5 cm diameter;; Large=3.0 cm diameter.

Adapted from Jani, S.K., Pennington, E.C., Wacha, J.E., Vigliotti, A.P., and Hussey, D.H., Correlation of point doses with total activity in intracavitary Cesium-137 applications for treating gynecologic cancers, *Endocurie. Hypertherm. Oncol.*, 4, 107, 1988.

## TABLE 4.5
### Dose Rates in cGy/hr in Water Around Ir-192 Wire Source
### (Linear activity = 1.0 mCi/cm; source diameter = 0.3 mm)

| Source Length (cm) | Distance From Source (cm) | | | | | | | | | | | | |
|---|---|---|---|---|---|---|---|---|---|---|---|---|---|
| | 0.2 | 0.3 | 0.4 | 0.5 | 0.6 | 0.7 | 0.8 | 0.9 | 1.0 | 2.0 | 3.0 | 4.0 | 5.0 |
| 1 | 56.1 | 32.7 | 21.4 | 15.1 | 11.0 | 8.5 | 6.7 | 5.4 | 4.5 | 1.2 | 0.5 | 0.3 | 0.2 |
| 2 | 63.3 | 39.8 | 28.0 | 21.0 | 16.3 | 13.1 | 10.7 | 9.0 | 7.5 | 2.2 | 1.0 | 0.6 | 0.4 |
| 3 | 65.5 | 42.2 | 30.5 | 23.5 | 18.7 | 15.4 | 12.8 | 10.9 | 9.4 | 3.1 | 1.5 | 0.9 | 0.5 |
| 4 | 66.3 | 43.2 | 31.7 | 24.7 | 19.9 | 16.6 | 14.0 | 12.1 | 10.5 | 3.8 | 1.9 | 1.1 | 0.7 |
| 5 | 66.8 | 44.1 | 32.3 | 25.3 | 20.7 | 17.3 | 14.8 | 12.8 | 11.3 | 4.3 | 2.2 | 1.3 | 0.9 |
| 6 | 67.2 | 44.1 | 32.7 | 25.8 | 21.1 | 17.8 | 15.3 | 13.3 | 11.7 | 4.7 | 2.5 | 1.5 | 1.0 |
| 7 | 67.6 | 44.5 | 33.0 | 26.1 | 21.4 | 18.1 | 15.6 | 13.7 | 12.1 | 5.0 | 2.7 | 1.7 | 1.2 |
| 8 | 67.6 | 44.5 | 33.2 | 26.3 | 21.7 | 18.4 | 15.8 | 13.9 | 12.3 | 5.3 | 2.9 | 1.9 | 1.3 |
| 10 | 67.6 | 44.9 | 33.5 | 26.6 | 22.0 | 18.7 | 16.2 | 14.3 | 12.7 | 5.6 | 3.3 | 2.1 | 1.5 |
| 12 | 68.1 | 44.9 | 33.6 | 26.8 | 22.1 | 18.9 | 16.4 | 14.5 | 12.9 | 5.9 | 3.5 | 2.3 | 1.6 |
| 14 | 68.5 | 45.4 | 33.7 | 26.9 | 22.3 | 19.0 | 16.5 | 14.6 | 13.1 | 6.0 | 3.6 | 2.5 | 1.8 |

From Kline, R.W., Gillin, M.T., Grimm, D.F., and Niroomand-Rad, A., Computer dosimetry of $^{192}$Ir wire, *Med. Phys.*, 12, 634, 1985. With permission.

TABLE 4.6
Decay Chart for Commonly Used Radioisotopes

| Time (days) | Physical Decay Factor | | | | | | | |
|---|---|---|---|---|---|---|---|---|
| | Au-198 | I-125 | I-131 | Ir-192 | P-32 | Pd-103 | Sm-145 | Yb-169 |
| 0 | 1.000 | 1.000 | 1.000 | 1.000 | 1.000 | 1.000 | 1.000 | 1.000 |
| 1 | 0.773 | 0.988 | 0.917 | 0.991 | 0.953 | 0.960 | 0.998 | 0.979 |
| 2 | 0.598 | 0.977 | 0.842 | 0.981 | 0.908 | 0.922 | 0.996 | 0.958 |
| 3 | 0.463 | 0.966 | 0.772 | 0.972 | 0.865 | 0.885 | 0.994 | 0.937 |
| 4 | 0.358 | 0.954 | 0.709 | 0.963 | 0.824 | 0.849 | 0.992 | 0.917 |
| 5 | 0.277 | 0.943 | 0.650 | 0.954 | 0.785 | 0.816 | 0.990 | 0.897 |
| 6 | 0.214 | 0.933 | 0.597 | 0.946 | 0.748 | 0.783 | 0.988 | 0.878 |
| 7 | 0.166 | 0.922 | 0.547 | 0.937 | 0.712 | 0.752 | 0.986 | 0.859 |
| 8 | 0.128 | 0.911 | 0.502 | 0.928 | 0.679 | 0.722 | 0.984 | 0.841 |
| 9 | 0.099 | 0.901 | 0.461 | 0.919 | 0.646 | 0.693 | 0.982 | 0.823 |
| 10 | 0.076 | 0.890 | 0.423 | 0.911 | 0.616 | 0.665 | 0.980 | 0.805 |
| 12 | 0.046 | 0.870 | 0.356 | 0.894 | 0.559 | 0.613 | 0.976 | 0.771 |
| 14 | 0.027 | 0.850 | 0.300 | 0.877 | 0.507 | 0.565 | 0.972 | 0.738 |
| 16 | 0.016 | 0.830 | 0.252 | 0.861 | 0.460 | 0.521 | 0.968 | 0.707 |
| 18 | 0.010 | 0.811 | 0.212 | 0.845 | 0.418 | 0.480 | 0.964 | 0.677 |
| 20 | 0.006 | 0.792 | 0.179 | 0.830 | 0.379 | 0.442 | 0.960 | 0.648 |
| 25 | — | 0.748 | 0.116 | 0.792 | 0.298 | 0.361 | 0.950 | 0.582 |
| 30 | — | 0.706 | 0.076 | 0.756 | 0.234 | 0.294 | 0.941 | 0.522 |
| 35 | — | 0.666 | 0.049 | 0.721 | 0.183 | 0.240 | 0.931 | 0.469 |
| 40 | — | 0.628 | 0.032 | 0.688 | 0.144 | 0.196 | 0.922 | 0.420 |
| 50 | — | 0.559 | 0.013 | 0.627 | 0.089 | 0.130 | 0.903 | 0.339 |
| 60 | — | 0.498 | 0.006 | 0.571 | 0.055 | 0.087 | 0.885 | 0.273 |
| 70 | — | 0.443 | — | 0.520 | 0.034 | 0.058 | 0.867 | 0.220 |
| 80 | — | 0.395 | — | 0.474 | 0.021 | 0.038 | 0.849 | 0.177 |
| 90 | — | 0.351 | — | 0.431 | 0.013 | 0.026 | 0.832 | 0.142 |
| 100 | — | 0.313 | — | 0.393 | 0.008 | 0.017 | 0.816 | 0.115 |

# REFERENCES

## GENERAL REFERENCES ON BRACHYTHERAPY

1. *A Handbook of Radioactivity Measurements Procedures*, NCRP Report No. 58, National Council on Radiation Protection and Measurements, Washington, D.C., 1978.

2. *Advances in Radiation Oncology Physics: Dosimetry, Treatment Planning, and Brachytherapy*, Medical Physics Monograph No. 19, Purdy, J.A., Ed., American Institute of Physics, Inc., New York, 1992.

3. *Dose and Volume Specification for Reporting Intracavitary Therapy in Gynecology*, ICRU Report 38, International Commission on Radiation Units and Measurements, Bethesda, MD, 1985.

4. *Handbook of Radioactive Nuclides*, Wang, Y., Ed., The Chemical Rubber Co., Cleveland, OH, 1969.

5. *Interstitial Brachytherapy: Physical, Biological and Clinical Considerations*, Interstitial Collaborative Working Group, Raven Press, New York, 1990.

6. *Modern Brachytherapy*, Pierquin, B., Wilson, J.F., and Chassagne, D., Eds., Masson Publishing USA, Inc., New York, 1987.

7. *Radiation Quantities and Units*, International Commission on Radiation Units and Measurements Report, Bethesda, MD, 1980.

8. *Radionuclides in Brachytherapy: Radium and After*, Trott, N.G., Ed., The British Institute of Radiology Suppl. 21, London, 1987.

9. *Recent Advances in Brachytherapy Physics*, AAPM Medical Physics Monograph No. 7, Shearer, D.R., Ed., American Institute of Physics, New York, 1981.

10. **Sakelliou, L., Sakellariou, K., Sarigiannis, K., Angelopoulos, A., Perris, A., and Zarris, G.**, Dose rate distributions around $^{60}$Co, $^{137}$Cs, $^{198}$Au, $^{192}$Ir, $^{241}$Am, $^{125}$I (models 6702 and 6711) brachytherapy sources and the nuclide $^{99}$Tc$^{m}$, *Phys. Med. Biol.*, 37, 1859, 1992.

11. **Shalek, R.J. and Stovall, M.**, Dosimetry in Implant Therapy, *Radiation Dosimetry, Vol. III*, Attix, F.H. and Tochilin, E., Academic Press, New York, 1969, 743.

12. *Specification of Brachytherapy Source Strength*, AAPM Report No. 21, American Institute of Physics, Inc., New York, 1987.

13. *Specification of Gamma-Ray Brachytherapy Sources*, NCRP Report No. 41, National Council on Radiation Protection and Measurements, Washington, D.C., 1974.

## A. DOSIMETRY OF AM-241 SOURCES

14. **Meigooni, A.S. and Nath, R.**, Tissue inhomogeneity correction for brachytherapy sources in a heterogeneous phantom with cylindrical symmetry, *Med. Phys.*, 19, 401, 1992.

15. **Muench, P.J. and Nath, R.**, Dose distributions produced by shielded applicators using $^{241}$Am for intracavitary irradiation of tumors in the vagina, *Med. Phys.*, 19, 1299, 1992.

16. **Nath, R. and Gray, L.**, Dosimetry studies on prototype [241]Am sources for brachytherapy, *Int. J. Radiat. Oncol., Biol. Phys.*, 13, 897, 1987.

17. **Nath, R., Gray, L., and Park, C.H.**, Dose distributions around cylindrical [241]Am sources for a clinical intracavitary applicator, *Med. Phys.*, 14, 809, 1987.

18. **Nath, R., Park, C.H., King, C., and Muench, P.**, A dose computation model for [241]Am vaginal applicators including the source-to-source shielding effect, *Med. Phys.*, 17, 833, 1990.

19. **Nath, R., Peschel, R.E., Park, C.H., and Fischer, J.J.**, Development of an [241]Am applicator for intracavitary irradiation of gynecologic cancers, *Int. J. Radiat. Oncol. Biol. Phys.*, 14, 969, 1988.

20. **Peschel, R.E., Dowling, S., Nath, R., Kacinski, B., Fischer, J.J., Chambers, J.T., Chambers, S.K., Kohorn, E.I., and Schwartz, P.E.**, An intracavitary vaginal applicator using americium-241, *Endocurie. Hypertherm. Oncol.*, 4, 91, 1988.

## B.   DOSIMETRY OF AU-198 SOURCES

21. **Akutsu, T., Watarai, J., Komatani, A., and Yamaguchi, K.**, Fundamental dose distribution for interstitial irradiation with Au grains, *Radiat. Med.*, 3, 234, 1985.

22. **Bulski, W., Danczak, Z., and Majenka, I.**, Computer dosimetry for interstitial therapy with [192]Ir and [198]Au implants, *Radiobiol. Radiother.*, 18, 447, 1977.

23. **Chenery, S.G., Japp, B., and Fitzpatrick, P.J.**, Dosimetry of radioactive gold grains for the treatment of choroidal melanoma, *Br. J. Radiol.*, 56, 415, 1983.

24. **Horiuchi, J., Takeda, M., Shibuya, H., Matsumoto, S., Hoshina, M., and Suzuki, S.**, Usefulness of [198]Au grain implants in the treatment of oral and oropharyngeal cancer, *Radiother. Oncol.*, 21, 29, 1991.

25. **Jani, S.K., Pennington, E.C., and Knosp, B.M.**, Dose anisotropy around an Au-198 seed source, *Med. Phys.*, 16, 632, 1989.

26. **Ling, C.C., Gromadzki, Z.C., Rustgi, S.N., and Cundiff, J.H.**, Directional dependence of radiation fluence from [192]Ir and [198]Au sources, *Radiol.*, 146, 791, 1983.

27. **Ling, C.C.**, Permanent implants using Au-198, Pd-103 and I-125: radiobiological considerations based on the linear quadratic model, *Int. J. Radiat. Oncol. Biol. Phys.*, 23, 81, 1992.

28. **Sharma, S.C.**, Procedures for interstitial radioactive gold grains, *Med. Dosim.*, 14, 23, 1989.

## C.   DOSIMETRY OF CF-252 SOURCES

29. **Beach, J.L. and Maruyama, Y.**, A simple method using film monitoring for CF-252 brachytherapy: technical note, *Radiat. Med.*, 2, 140, 1984.

30. **Beach, J.L., Schroy, C.B., Ashtari, M., Harris, M.R., and Maruyama, Y.**, Boron neutron capture enhancement of [252]Cf brachytherapy, *Int. J. Radiat. Oncol. Biol. Phys.*, 18, 1421, 1990.

31. **Iyer, P.S., Gopalakrishnan, A.K., and Venkataraman, G.**, Californium-252 line sources: experimental verification of depth dose data using NTA films and dosage tables based on Paterson and Parker rules, *Phys. Med. Biol.*, 21, 98, 1976.

32. **Jones, T.D. and Auxier, J.A.**, Local dose from neutron-produced-recoil ions in the region of a therapeutic $^{252}$Cf needle, *Radiol.*, 104, 187, 1972.

33. **Krishnaswamy, V.**, Calculation of the dose distribution about californium-252 needles in tissue, *Radiol.*, 98, 155, 1971.

34. **Krishnaswamy, V.**, Linear energy transfer distributions for neutrons about a $^{252}$Cf point source in tissue, *Radiol.*, 101, 417, 1971.

35. **Maruyama, Y., Beach, J.L., Hazle, J., Ashtari, M., Schroy, C.B., and Olson, M.H.**, Therapeutic dosimetry for Cf-252 neutron brachytherapy of pelvic cancer, *Int. J. Radiat. Oncol. Biol. Phys.*, 11, 927, 1985.

36. **Maruyama, Y., Tai, D.L., and Beach, J.L.**, Neutron and gamma dose distribution for tandem, tandem and ovoid and plaque for intracavitary $^{252}$Cf therapy of cervix carcinoma, *Radiat. Med.*, 1, 230, 1983.

37. **Oliver, Jr., G.D. and Wright, C.N.**, Dosimetry of an implantable $^{252}$Cf source, *Radiol.*, 92, 143, 1969.

38. **Paine, C.H., Berry, R.J., Stedeford, J.B., Barber, C.D., Young, C.M., and Wiernik, G.**, The use of californium-252 sources for brachytherapy of human tumors: a preliminary report, *Eur. J. Cancer*, 10, 365, 1974.

39. **Permar, D.H.**, Cf-252 neutron sources for interstitial afterloading, *Int. J. Radiat. Oncol. Biol. Phys.*, 1, 1003, 1976.

40. **Robertson, J.S., Fairchild, R.G. and Atkins, H.L.**, Dosimetry of californium 252, *Radiol.*, 104, 393, 1972.

41. **Smith, A., Almond, P., and Delclos, L.**, Evaluations of $^{252}$Cf neutron emitter for interstitial and intracavitary radiation therapy, *Eur. J. Cancer*, 10, 369, 1974.

42. **Stoddard, D.H.**, Calculated dose rate from $^{252}$Cf radiotherapy source, *Radiol.*, 103, 187, 1972.

43. **Tai, D.L. and Maruyama, Y.**, Assessment of neutron/gamma-ray dose ratios in intracavitary $^{252}$Cf neutron therapy, *Radiol.*, 128, 795, 1978.

44. **Veerling, J.P., Oliver, Jr., G.D., and Moore, E.B.**, A storage and handling facility for californium-252 medical sources, *Health Physics*, 25, 163, 1973.

45. **Wright, C.N., Bonlogne, A.R., Reinig, W.C., and Evans, A.G.**, Implantable Californium-252 neutron sources for radiotherapy, *Radiol.*, 89, 337, 1967.

## D. DOSIMETRY OF CS-137 SOURCES

46. **Breitman, K.E.**, Dose-rate tables for clinical $^{137}$Cs sources sheathed in platinum, *Br. J. Radiol.*, 47, 657, 1974.

47. **Dale, R.G.**, Some theoretical derivations relating to the tissue dosimetry of brachytherapy nuclides, with particular reference to iodine-125, *Med. Phys.*, 10, 176, 1983.

48. **Diffey, B.L. and Klevenhagen, S.C.**, An experimental and calculated dose distribution in water around CDC K-type Caesium-137 sources, *Phys. Med. Biol.*, 20, 446, 1975.

49. **Horsler, A.F.C., Jones, J.C., and Stacey, A.J.**, Caesium 137 sources for use in intracavitary and interstitial radiotherapy, *Br. J. Radiol.*, 37, 385, 1964.

50. **Jani, S.K., Pennington, E.C., Wacha, J.E., Vigliotti, A.P., and Hussey, D.H.**, Correlation of point doses with total activity in intracavitary Cesium-137 applications for treating gynecologic cancers, *Endocurie. Hypertherm. Oncol.*, 4, 107, 1988.

51. **Klevenhagen, S.C.**, An experimental study of the dose distribution in water around $^{137}$Cs tubes used in brachytherapy, *Br. J. Radiol.*, 46, 1073, 1973.

52. **Krishnaswamy, V.**, Dose distributions about $^{137}$Cs sources in tissue, *Radiol.*, 105, 181, 1972.

53. **Meli, J.A., Meigooni, A.S., and Nath, R.**, Comments on "Radial dose distribution of $^{192}$Ir and $^{137}$Cs seed sources", *Med. Phys.,* 16, 824, 1989.

54. **Prasad, S.C., Bassano, D.A., and Kubsad, S.S.**, Buildup factors and dose around a $^{137}$Cs source in the presence of inhomogeneities, *Med. Phys.*, 10, 705, 1983.

55. **Rossiter, M.J., Williams, T.T., and Bass, G.A.**, Air kerma rate calibration of small sources of $^{60}$Co, $^{137}$Cs, $^{226}$Ra and $^{192}$Ir, *Phys. Med. Biol.*, 36, 279, 1991.

56. **Sharma, S.C., Gerbi, B., and Madoc-Jones, H.**, Dose rates for brachytherapy applicators using $^{137}$Cs sources, *Int. J. Radiat. Oncol. Biol. Phys.*, 5, 1893, 1979.

57. **Thomason, C. and Higgins, P.**, Erratum: radial dose distribution of $^{192}$Ir and $^{137}$Cs seed sources [Med. Phys. 16, 254, (1989)], *Med. Phys.*, 16, 826, 1989.

58. **Thomason, C. and Higgins, P.**, Radial dose distribution of $^{192}$Ir and $^{137}$Cs seed sources, *Med. Phys.*, 16, 254, 1989.

59. **Thomason, C., Mackie, T.R., and Lindstrom, M.J.**, Effect of source encapsulation on the energy spectra of $^{192}$Ir and $^{137}$Cs seed sources, *Phys. Med. Biol.*, 36, 495, 1991.

60. **Thomason, C., Mackie, T.R., Lindstrom, M.J., and Higgins, P.D.**, The dose distribution surrounding $^{192}$Ir and $^{137}$Cs seed sources, *Phys. Med. Biol.*, 36, 475, 1991.

61. **Williamson, J.F.**, Monte Carlo and analytic calculation of absorbed dose near $^{137}$Cs intracavitary sources, *Int. J. Radiat. Oncol. Biol. Phys.*, 15, 227, 1988.

## E.    DOSIMETRY OF I-125 SOURCES

62. **Ahmad, M., Fontenla, D.P., Chiu-Tsao, S-T., Chui, C.S., Reiff, J.E., Anderson, L.L., Huang, D.Y.C., and Schell, M.C.**, Diode dosimetry of models 6711 and 6712 $^{125}$I seeds in a water phantom, *Med. Phys.*, 19, 391, 1992.

63. **Anderson, L.L., Kuan, H.M., and Ding, I-Y.**, Clinical dosimetry with $^{125}$I, in *Modern Interstitial and Intracavitary Radiation Management*, George III, F.W., Ed., Masson Publishing USA, Inc., New York, 1981, 9.

64. **Burns, G.S. and Raeside, D.E.**, The accuracy of single-seed dose superposition for I-125 implants, *Med. Phys.*, 16, 627, 1989.

65. **Burns, G.S. and Raeside, D.E.**, Two-dimensional dose distribution around a commercial $^{125}$I seed, *Med. Phys.*, 15, 56, 1988.

66. **Chiu-Tsao, S-T., Anderson, L.L., O'Brien, K., and Sanna, R.**, Dose rate determination for $^{125}$I seeds, *Med. Phys.*, 17, 815, 1990.

67. **Chiu-Tsao, S-T., O'Brien, K., Sanna, R., Tsao, H-S., Vialotti, C., Chang, Y.-S., Rotman, M., and Packer, S.**, Monte Carlo dosimetry for $^{125}$I and $^{60}$Co in eye plaque therapy, *Med. Phys.*, 13, 678, 1986.

68. **Cygler, J., Szanto, J., and Soubra, M.**, Effects of gold and silver backings on the dose rate around an $^{125}$I seed, *Med. Phys.*, 17, 172, 1990.

69. **Dale, R.G.**, Some theoretical derivations relating to the tissue dosimetry of brachytherapy nuclides, with particular reference to iodine-125, *Med. Phys.*, 10, 176, 1983.

70. **Hartmann, G.H., Schlegel, W., and Scharfenberg, H.**, The three-dimensional dose distribution of $^{125}$I seeds in tissue, *Phys. Med. Biol.*, 28, 693, 1983.

71. **Hashemi, A.M., Mills, M.D., Hogstrom, K.R., and Almond, P.R.**, The exposure rate constant for a silver wire $^{125}$I seed, *Med. Phys.*, 15, 228, 1988.

72. **Jani, S.K., Hitchon, P.W., VanGilder, J.C., Pennington, E.C., and Hussey, D.H.**, Choice of radioisotope in stereotactic interstitial radiotherapy of small brain tumors, *Appl. Neurophysiol.*, 50, 295, 1987.

73. **Krishnaswamy, V.**, Dose distribution around an $^{125}$I seed source in tissue, *Radiol.*, 126, 489, 1978.

74. **Krishnaswamy, V.**, Dose tables for $^{125}$I seed implants, *Radiol.*, 132, 727, 1979.

75. **Kubo, H.**, Comparison of two independent exposure measurement techniques for clinical I-125 seeds, *Med. Phys.*, 12, 221, 1985.

76. **Kubo, H.**, Determination of the half-life of I-125 encapsulated clinical seeds during a Si(Li) detector, *Med. Phys.*, 10, 889, 1983.

77. **Ling, C.C. and Yorke, E.D.**, Interface dosimetry for I-125 sources, *Med. Phys.*, 16, 376, 1989.

78. **Ling, C.C., Anderson, L.L., and Shipley, W.U.**, Dose inhomogeneity in interstitial implants using $^{125}$I seeds, *Int. J. Radiat. Oncol. Biol. Phys.*, 5, 419, 1979.

79. **Ling, C.C., Gromadzki, Z.C., Rustgi, S.N., and Cundiff, J.H.**, Directional dependence of radiation fluence from Ir-192 and Au-198 sources, *Radiol.*, 146, 791, 1983.

80. **Ling, C.C., Schell, M.C., Yorke, E.D., Palos, B.B., and Kubiatowicz, D.O.**, Two-dimensional dose distribution of $^{125}$I seeds, *Med. Phys.*, 12, 652, 1985.

81. **Ling, C.C., Yorke, E.D., Spiro, I.J., Kubiatowicz, D., and Bennett, D.**, Physical dosimetry of $^{125}$I seeds of a new design for interstitial implant, *Int. J. Radiat. Oncol. Biol. Phys.*, 9, 1747, 1983.

82. **Luxton, G., Astrahan, M.A., and Petrovich, Z.**, Backscatter measurements from a single seed of $^{125}$I for ophthalmic plaque dosimetry, *Med. Phys.*, 15, 397, 1988.

83. **Luxton, G., Astrahan, M.A., Findley, D.O., and Petrovich, A.**, Measurement of dose rate from exposure-calibrated $^{125}$I seeds, *Int. J. Radiat. Oncol. Biol. Phys.*, 18, 1199, 1990.

84. **Luxton, G., Astrahan, M.A., Liggett, P.E., Neblett, D.L., Cohen, D.M., and Petrovich, Z.**, Dosimetric calculations and measurements of gold plaque ophthalmic irradiators using iridium-192 and iodine-125 seeds, *Int. J. Radiat. Oncol. Biol. Phys.*, 15, 167, 1988.

85. **Meigooni, A.S. and Nath, R.**, Tissue inhomogeneity correction for brachytherapy sources in a heterogeneous phantom with cylindrical symmetry, *Med. Phys.*, 19, 401, 1992.

86. **Meigooni, A.S., Meli, J.A., and Nath, R.**, A comparison of solid phantoms with water for dosimetry of $^{125}$I brachytherapy sources, *Med. Phys.*, 15, 695, 1988.

87. **Meigooni, A.S., Meli, J.A., and Nath, R.**, Interseed effects on dose for $^{125}$I brachytherapy implants, *Med. Phys.*, 19, 385, 1992.

88. **Nath, R., Meigooni, A.S., and Meli, J.A.**, Dosimetry on transverse axes of $^{125}$I and $^{192}$Ir interstitial brachytherapy sources, *Med. Phys.*, 17, 1032, 1990.

89. **Nikesch, W.**, Problems with the radiation measurements incorporated/American College of Radiology mammography accreditation phantom, *Med. Phys.*, 17, 934, 1990.

90. **Prasad, S.C., Bassano, D.A., and Fear, P.I.**, Dose distribution for $^{125}$I implants due to anisotropic radiation emission and unknown seed orientation, *Med. Phys.*, 14, 296, 1987.

91. **Rao, G.U.V., Kan, P.T., and Howells, R.**, Interstitial volume implants with I-125 seeds, *Int. J. Radiat. Oncol. Biol. Phys.*, 7, 431, 1981.

92. **Rosenow, U.F., Findlay, P.A., and Wright, D.C.**, The NCI-atlas of dose distributions for regular $^{125}$I brain implants, *Radiother. Oncol.*, 10, 127, 1987.

93. **Rustgi, S.N.**, Photon spectral characteristics of a new double-walled iodine-125 source, *Med. Phys.*, 19, 927, 1992.

94. **Scarbrough, E.C., Sanborn, G.E., Anderson, J.A., Nguyen, P.D., Niederkorn, J.Y., and Antich, P.P.**, Dose distribution around a 3.0-mm type 6702 I-125 seed, *Med. Phys.*, 17, 460, 1990.

95. **Schell, M.C., Ling, C.C., Gromadzki, Z.C., and Working, K.R.**, Dose distributions of model 6702 I-125 seeds in water, *Int. J. Radiat. Oncol. Biol. Phys.*, 13, 795, 1987.

96. **Sondhaus, C.A.**, $^{125}$I: Physical properties, photon dosimetry, and effectiveness, in *Modern Interstitial and Intracavitary Radiation Management*, George III, F.W., Masson Publishing USA, Inc., New York, 1981, 83.

97. **Weaver, K.A., Smith, V., Huang, D., Barnett, C., Schell, M.C., and Ling, C.**, Dose parameters of $^{125}$I and $^{192}$Ir seed sources, *Med. Phys.*, 16, 636, 1989.

98. **Weaver, K.A., Smith, V., Huang, D., Barnett, C., Schell, M.C., and Ling, C.,** Dose parameters of $^{125}$I and $^{192}$Ir seed sources, *Med. Phys.*, 16, 636, 1989.

99. **Williamson, J.F. and Quintero, F.J.,** Theoretical evaluation of dose distributions in water about models 6711 and 6702 $^{125}$I seeds, *Med. Phys.*, 15, 891, 1988.

100. **Williamson, J.F.,** Comparison of measured and calculated dose rates in water near I-125 and Ir-192 seeds, *Med. Phys.*, 18, 776, 1991.

101. **Wu, A., Sternick, E.S., and Muise, D.J.,** Effect of gold shielding on the dosimetry of an $^{125}$I seed at close range, *Med. Phys.*, 15, 627, 1988.

102. **Yorke, E.D., Huang, Y.C.D., Schell, M.C., Wong, R., and Ling, C.C.,** Clinical implications of I-125 dosimetry of bone and bone-soft tissue interfaces, *Int. J. Radiat. Oncol. Biol. Phys.*, 21, 1613, 1991.

## F. DOSIMETRY OF I-131 RADIONUCLIDE

103. **Beierwaltes, W.H.** The treatment of hyperthyroidism with I-131, *Semin. Nucl. Med.*, 8, 95, 1978.

104. **Beierwaltes, W.H.,** The treatment of thyroid carcinoma with radioactive iodine, *Semin. Nucl. Med.*, 8, 79, 1978.

105. **Benua, R.S., Cicale, N., Sonenberg, M., and Rawson, R.W.,** The relation of radioiodine dosimetry to results and complications in the treatment of metastatic thyroid cancer, *Am. J. Roent. Rad. Ther. Nucl. Med.*, 87, 171, 1962.

106. **Maxon, III, H.R., Englaro, E.E., Thomas, S.R., Hertzberg, V.S., Hinnefeld, J.D., Chen, L.S., Smith, H., Cummings, D., and Aden, M.D.,** Radioiodine-131 therapy for well-differentiated thyroid cancer—a quantitative radiation dosimetric approach: outcome and validation in 85 patients, *J. Nucl. Med.*, 33, 1132, 1992.

107. **O'Connell, M.E.A., Steere, H.A., and Trott, N.G.,** The application of a digital whole body scanner to the dosimetry of intralymphatic $^{32}$P/$^{131}$I lipiodol, *Br. J. Radiol.*, 49, 779, 1976.

108. **Ponto, J.A. and Chilton, H.M.,** Therapeutic applications of radiopharmaceuticals: thyroid disease, polycythemia vera, and malignant effusion, in *Pharmaceuticals in Medical Imaging*, Swanson, D.P., Chilton, H.M., and Thrall, J.H., Eds., MacMillan Publishing Co., New York, 1990, 599.

109. **Schlesinger, T., Flower, M.A., and McCready, V.R.,** Radiation dose assessments in radioiodine (I131) therapy. I. The necessity for in vivo quantitation and dosimetry in the treatment of carcinoma of the thyroid, *Radiother. Oncol.*, 14, 35, 1989.

110. **St. Germain, J. and Finn, R.D.,** I-131 and P-32 radiopharmaceuticals: procedures, radiation protection and quality assurance, AAPM, 1990, 1.

## G. DOSIMETRY OF IR-192 SOURCES

111. **Abrath, F.G., Henderson S.D., Simpson, J.R., Moran, C.J., and Marchosky, J.A.**, Dosimetry of CT-guided volumetric Ir-192 brain implant, *Int. J. Radiat. Oncol. Biol. Phys.*, 12, 359, 1986.

112. **Boyer, A.L. and Cobb, P.D.**, [192]Ir Hospital calibration procedures, in *Recent Advances in Brachytherapy Physics*, Shearer, D.R., Ed., AAPM Medical Physics Monograph No. 7 , American Institute of Physics, New York, 1979, 82.

113. **Dale, R.G.**, Some theoretical derivations relating to the tissue dosimetry of brachytherapy nuclides, with particular reference to iodine-125, *Med. Phys.*, 10, 176, 1983.

114. **Findley, D.O. and Forell, B.W.**, Routine quality control of Ir-192 sources: results of 46 months of surveillance, *Int. J. Radiat. Oncol. Biol. Phys.*, 11, 1727, 1985.

115. **Gillin, M.T., Lopez, F., Kline, R.W., Grimm, D.F., and Niroomand-Rad, A.**, Comparison of measured and calculated dose distributions around an iridium-192 wire, *Med. Phys.*, 15, 915, 1988.

116. **Glasgow, G.P. and Dillman, L.T.**, Specific γ-ray constant and exposure rate constant of [192]Ir, *Med. Phys.*, 6, 49, 1979.

117. **Glasgow, G.P. and Dillman, L.T.**, Specific gamma-ray constant and exposure rate constant of [192]Ir, *Med. Phys.*, 6, 49, 1979.

118. **Glasgow, G.P.**, Exposure rate constants for filtered [192]Ir sources, *Med. Phys.*, 8, 502, 1981.

119. **Glasgow, G.P.**, The ratio of the dose in water and exposure in air for a point source of [192]Ir: a review in recent advances in brachytherapy physics, Shearer, D.R., Ed., Medical Physics Monograph No. 7 (AAPM), New York, 1987, 104.

120. **Goetsch, S.J., Attix, F.H., Pearson, D.W., and Thomadsen, B.R.**, Calibration of [192]Ir high-dose-rate afterloading systems, *Med. Phys.*, 18, 462, 1991.

121. **Jani, S.K., Pennington, E.C., Knosp, B.M., and Doornbos, J.F.**, Analysis of relative dose distribution around iridium-192 endobronchial implants, *Endocurie. Hypertherm. Oncol.*, 5, 187, 1989.

122. **Kline, R.W., Gillin, M.T., Grimm, D.F., and Niroomand-Rad, A.**, Computer dosimetry of [192]Ir wire, *Med. Phys.*, 12, 634, 1985.

123. **Ling, C.C., Gromadzki, Z.C., Rustgi, S.N., and Cundiff, J.H.**, Directional dependence of radiation fluence from [192]Ir and [198]Au sources, *Radiol.*, 146, 791, 1983.

124. **Loftus, T.P.**, Standardization of iridium-192 gamma ray sources in terms of exposure, *J. Res. Natl. Bur. Stand.*, 85, 19, 1979.

125. **Luxton, G., Astrahan, M.A., Liggett, P.E., Neblett, D.L., Cohen, D.M., and Petrovich, Z.**, Dosimetric calculations and measurements of gold plaque ophthalmic irradiators using iridium-192 and iodine-125 seeds, *Int. J. Radiat. Oncol. Biol. Phys.*, 15, 167, 1988.

126. **Marinello, G., Valero, M., Leung, S., and Pierquin, B.**, Comparative dosimetry between iridium wires and seed ribbons, *Int. J. Radiat. Oncol. Biol. Phys.*, 11, 1733, 1985.

127. **Marinello, G., Valero, M., Leung, S., and Pierquin, B.,** Comparative dosimetry between iridium wires and seed ribbons, *Int. J. Radiat. Oncol. Biol. Phys.,* 11, 1733, 1985.

128. **Marks, J.E., Oliver, G.D., and Velkley, D.,** A method of increasing the linear activity of $^{192}$Ir sources for interstitial implantation, *Radiol.,* 128, 511, 1978.

129. **Meigooni, A.S., Meli, J.A., and Nath, R.,** Influence of the variation of energy spectra with depth in the dosimetry of $^{192}$Ir using LiF TLD, *Phys. Med. Biol.,* 33, 1159, 1988.

130. **Meisberger, L.L., Keller, R.J., and Shalek, R.J.,** The effective attenuation in water of the gamma rays of gold-198, iridium-192, cesium-137, radium-226 and cobalt-60, *Radiat.,* 90, 953, 1968.

131. **Meli, J.A., Meigooni, A.S., and Nath, R.,** Comments on "Radial dose distribution of $^{192}$Ir and $^{137}$Cs seed sources," by C. Thomason and P. Higgins, letter to *Med. Phys.,* 16, 824, 1989.

132. **Meli, J.A., Meigooni, A.S., and Nath, R.,** On the choice of phantom material for the dosimetry of $^{192}$Ir sources, *Int. J. Radiat. Oncol. Biol. Phys.,* 14, 587, 1988.

133. **Meli, J.A., Meigooni, A.S., and Nath, R.,** On the choice of phantom material for the dosimetry of $^{192}$Ir sources, *Int. J. Radiat. Oncol. Biol. Phys.,* 14, 587, 1988.

134. **Meli, J.A., Meigooni, A.S., and Nath, R.,** Rebuttal to the letter-to-the-editor of R.G. Dale, *Int. J. Radiat. Oncol. Biol. Phys.,* 16, 1654, 1989.

135. **Muller-Runkel, R. and Vijayakumar, S.,** Brachytherapy implants with differently spaced Ir-192 seeds: a dosimetric study, *Radiol.,* 165, 271, 1987.

136. **Murphy, D.J., Memula, N., and Doss, L.L.,** A $^{192}$Ir nomogram system for single plane implants, *Int. J. Radiat. Oncol. Biol. Phys.,* 12, 267, 1986.

137. **Murphy, D.J., Memula, N., and Doss, L.L.,** A $^{192}$Ir nomogram system for single plane implants, *Int. J. Radiat. Oncol. Biol. Phys.,*12, 267, 1986.

138. **Murphy, Jr., D.J. and Doss, L.L.,** Small computer algorithms for comparing therapeutic performances of single-plane iridium implants, *Med. Phys.,* 11, 193, 1984.

139. **Nath, R., Meigooni, A.S., and Meli, J.A.,** Dosimetry on transverse axes of $^{125}$I and $^{192}$Ir interstitial brachytherapy sources, *Med. Phys.,* 17, 1032, 1990.

140. **Olch, A.J. and Kagan, A.R.,** Comments on "Reference dose rates for single- and double-plane $^{192}$Ir implants ", *Med. Phys.,* 16, 143, 1989.

141. **Rosen, I.I., Khan, K.M., Lane, R.G., and Kelsey, C.A.,** The effect of geometric errors in the reconstruction of iridium-192 seed implants, *Med. Phys.,* 9, 220, 1982.

142. **Rossiter, M.J., Williams, T.T., and Bass, G.A.,** Air kerma rate calibration of small sources of $^{60}$Co, $^{37}$Cs, $^{226}$Ra and $^{192}$Ir, *Phys. Med. Biol.,* 36, 279, 1991.

143. **Saw, C.B. and Suntharalingam, N.,** Reply to comments of Olch and Kagan, *Med. Phys.,* 16, 144, 1989.

144. **Thomason, C. and Higgins, P.,** Erratum: radiation dose distribution of $^{192}$Ir and $^{137}$Cs seed sources [Med. Phys. 16, 254 (1989)], *Med. Phys.,* 16, 826, 1989.

145. **Thomason, C. and Higgins, P.**, Radial dose distribution of [192]Ir and [137]Cs seed sources, *Med. Phys.*, 16, 254, 1989.
146. **Thomason, C., Mackie, T.R., and Lindstrom, M.J.**, Effect of source encapsulation on the energy spectra of [192]Ir and [137]Cs seed sources, *Phys. Med. Biol.*, 36, 495, 1991.
147. **Thomason, C., Mackie, T.R., Lindstrom, M.J., and Higgins, P.D.**, The dose distribution surrounding [192]Ir and [137]Cs seed sources, *Phys. Med. Biol.*, 36, 475, 1991.
148. **Thomason, C.**, Radial dose distribution from [192]Ir seeds at distances far from the source, *Med. Phys.*, 19, 199, 1992.
149. **Tripathi, U.B. and Shanta, A.**, A general formula for computation of tissue-air ratios for radionuclides commonly used in brachytherapy, *Med. Phys.*, 12, 88, 1985.
150. **Weaver, K.A., Smith, V., Huang, D., Barnett, C., Schell, M.C., and Ling, C.**, Dose parameters of [125]I and [192]Ir seed sources, *Med. Phys.*, 16, 636, 1989.
151. **Williamson, J.F.**, Comparison of measured and calculated dose rates in water near I-125 and Ir-192 seeds, *Med. Phys.*, 18, 776, 1991.
152. **Williamson, J.F.**, The accuracy of the line and point source approximations in [192]Ir dosimetry, *Int. J. Radiat. Oncol. Biol. Phys.*, 12, 409, 1986.

## H.  DOSIMETRY OF P-32 RADIONUCLIDE

153. **Ponto, J.A. and Chilton, H.M.**, Therapeutic applications of radiopharmaceuticals: thyroid disease, polycythemia vera, and malignant effusion, in *Pharmaceuticals in Medical Imaging*, Swanson, D.P., Chilton, H.M., and Thrall, J.H., Eds., MacMillan Publishing Co., New York, 1990, 599.
154. **Vynckier, S. and Wambersie, A.**, Dosimetry of beta sources in radiotherapy I. The beta point source dose function, *Phys. Med. Biol.*, 27, 1339, 1982.
155. **O'Connell, M.E.A., Steere, H.A., and Trott, N.G.**, The application of a digital whole body scanner to the dosimetry of intralymphatic [32]P/[131]I lipiodol, *Br. J. Radiol.*, 49, 779, 1976.
156. **Kaplan, W.D., Zimmerman, R.E., Bloomer, W.D., Knapp, R.C., and Adelstein, S.J.**, Therapeutic intraperitoneal [32]P: a clinical assessment of the dynamics of distribution, *Radiol.*, 138, 683, 1981.
157. **Taasan, V., Shapiro, B., Taren, J.A., Beierwaltes, W.H., McKeever, P., Wahl, R.L., Carey, J.E., Petry, N., and Mallette, S.**, Phosphorus-32 therapy of cystic grade IV astrocytomas: technique and preliminary application, *J. Nucl. Med.*, 26, 1335, 1985.
158. **St. Germain, J. and Finn, R.D.**, I-131 and P-32 radiopharmaceuticals: procedures, radiation protection and quality assurance, AAPM, 1990, 1.
159. **Nori, D. and Hilaris, B.**, Role of radiocolloids in gynecological cancer, in *Radiation Therapy in Gynecological Cancer*, Hilaris, B. and Nori, D., Eds., Alan R. Liss, Inc., New York, NY, 1987, 309.

## I. DOSIMETRY OF PD-103 SOURCES

160. **Chiu-Tsao, S-T., and Anderson, L.L.**, Thermoluminescent dosimetry for $^{103}$Pd seeds (model 200) in solid water phantom, *Med. Phys.*, 18, 449, 1991.

161. **Meigooni, A.S. and Nath, R.**, Tissue inhomogeneity correction for brachytherapy sources in a heterogeneous phantom with cylindrical symmetry, *Med. Phys.*, 19, 401, 1992.

162. **Meigooni, A.S., Sabnis, S., and Nath, R.**, Dosimetry of Palladium-103 brachytherapy sources for permanent implants, *Endocurie. Hypertherm. Oncol.*, 6, 107, 1990.

163. **Nath, R., Meigooni, A.S., and Melillo, A.**, Some treatment planning considerations for $^{103}$Pd and $^{125}$I permanent interstitial implants, *Int. J. Radiat. Oncol. Biol. Phys.*, 22, 1131, 1992.

164. **Russell, Jr., J.L.**, Calculated dose from TheraSeed implants, Theragenics Corporation Internal Report TM-10014C, November 4, 1984.

## J. DOSIMETRY OF RA-226 SOURCES

165. Brachytherapy—intercavitary and interstitial sources, in *The physics of radiology*, Johns, H.E. and Cunningham, J.R., Eds., Charles C. Thomas, Springfield, IL, 1983, 453.

166. **Gallaghar, R.G. and Saenger, E.L.**, Radium capsules and their associated hazards, *Amer. J. Roentgenol.*, 77, 511, 1957.

167. **Harrington, E.L.**, Radium: radon plants, in *Medical Physics*, Glasser, O., Ed., Chicago, 1944, 1193.

168. **Paterson, R. and Parker, H.M.**, A dosage system for interstitial radium therapy, *Br. J. Radiol.*, 11, 252, 1938.

169. **Paterson, R.**, A dosage system for gamma ray therapy, *Br. J. Radiol.*, 7, 592, 1934.

170. **Paterson, R., Parker, H.M., and Spiers, F.W.**, A system of dosage for cylindrical distributions of radium, *Br. J. Radiol.*, 9, 487, 1936.

171. **Quimby, E.H.**, Dosage tables for linear radium sources, *Radiol.*, 43, 572, 1944.

172. *Radium Dosage*, the Manchester System, Livingstone, E. and Livingstone, S., Eds., Edinburgh, 1967.

173. The Radium Dosage System, in *The Treatment of Malignant Disease by Radium and X-Rays*, Paterson, R., Ed., Edward Arnold, London, 1948, 133.

174. **Young, M.E.J. and Batho, H.F.**, Dose tables for linear radium sources calculated by an electronic computer, *Br. J. Radiol.*, 37, 38, 1964.

## K. DOSIMETRY OF SM-145 SOURCES

175. **Fairchild, R.G., Kalef-Ezra, J., Packer, S., Wielopolski, L., Laster, B.H., Robertson, J.S., Mausner, L., and Kanellitsas, C.**, Samarium-145: a new brachytherapy source, *Phys. Med. Biol.*, 32, 847, 1987.

176. **Meigooni, A.S. and Nath, R.**, A comparison of radial dose functions for $^{103}$Pd, $^{125}$I, $^{145}$Sm, $^{241}$Am, $^{169}$Yb, $^{192}$Ir, and $^{137}$Cs brachytherapy sources, *Int. J. Radiat. Oncol. Biol. Phys.*, 22, 1125, 1992.

## L. DOSIMETRY OF SR-90 SOURCES

177. **Ali, M.M. and Khan, F.M.**, Determination of surface dose rate from a $^{90}$Sr ophthalmic applicator, *Med. Phys.*, 17, 416, 1990.

178. **Coffey, C., Sayeg, J., Beach, L., Song, S., Landis, C., and Conners, A.**, Calibration of surface dose rate for a Sr-90 beta applicator-comparison of experimental, theoretical, and biological methods, *Med. Phys.*, 8, 558, 1981.

179. **Goetsch, S.J. and Sunderland, K.S.**, Surface dose rate calibration of Sr-90 plane ophthalmic applicators, *Med. Phys.*, 18, 161, 1991.

180. **Hendee, W.R.**, Measurement and correction of nonuniform surface dose rates from beta eye applicators, *Am. J. Roentgenol.*, 103, 734, 1968.

181. **Hendee, W.R.**, Thermoluminescent dosimetry of beta depth dose, *Am. J. Roentgenol.*, 97, 1046, 1966.

182. **Jones, C.H. and Dermentzoglou**, Practical aspects of $^{90}$Sr ophthalmic applicator dosimetry, *Br. J. Radiol.*, 44, 203, 1971.

183. **Kahlig, C.A., Smith, A.R., and Almond, P.**, Film calibration of a strontium-90 ophthalmic applicator, *Am. J. Roentgenol.*, 123, 36, 1975.

184. **Pruitt, J.S.**, Calibration of beta-particle ophthalmic applicators at the National Bureau of Standards, *J. Res. Nat. Bur. Stand.*, 91, 165, 1986.

185. **Pruitt, J.S.**, *Calibration of Beta-Particle-Emitting Ophthalmic Applicators*, U.S. Department of Commerce, NBS Special Publication 250-9 U.S. GPO, Washington, D.C., 1987.

186. **Reft, C.S., Kuchnir, F.T., Rosenberg, I., and Myrianthopoulos, L.C.**, Dosimetry of Sr-90 ophthalmic applicators, *Med. Phys.*, 17, 641, 1990.

187. **Ruden, B.I. and Bengtsson, G.**, TLD measurement of dose distribution around a beta-ray eye applicators, *Phys. Med. Biol.*, 19, 186, 1974.

188. **Sauer, R.**, Radiotherapy and cisplatin for invasive bladder cancer: criteria of "Complete Response", *Int. J. Radiat. Oncol. Biol. Phys.*, 16, 1653, 1989.

189. **Sayeg, J.A. and Gregory, R.C.**, A new method for characterizing beta-ray ophthalmic applicator sources, *Med. Phys.*, 18, 453, 1991.

190. **Sayeg, J.A.**, Radiochromic radiation detectors for the calibration of Sr$^{90}$ beta eye applicators, *Med. Phys.*, 10, 133, 1983.

191. **Sinclair, W.K. and Trott, N.G.**, The construction and measurement of beta-ray applicators for use in ophthalmology, *Br. J. Radiol.*, 29, 15, 1956.

192. **Supe, S.J. and Cunningham, J.R.**, A physical study of a strontium-90 beta-ray applicator, *Am. J. Roentgenol.*, 89, 570, 1963.

193. **Supe, S.J., Mallikarjuna, S., and Sawant, S.G.**, Dosimetry of spherical Sr$^{90}$-Y$^{90}$ beta-ray eye applicators, *Am. J. Roentgenol.*, 123, 26, 1975.

## M. DOSIMETRY OF YB-169

194. Battista, J.J., Porter, A.T., Mason, D., and Barnett, R., Ytterbium169 — a novel brachytherapy source, Radiat. Oncol. Biol. Phys., 19 (Suppl. 1), 152, 1990.

195. **Loft, S.M., Coles, I.P., and Dale, R.G.**, The potential of ytterbium 169 in brachytherapy: a brief physical and radiobiological assessment, *Br. J. Radiol.*, 65, 252, 1992.

196. **Mason, D.L.D., Battista, J.J., Barnett, R.B., and Porter, A.T.**, Ytterbium-169: calculated physical properties of a new radiation source for brachytherapy, *Med. Phys.*, 19, 695, 1992.

197. **Meigooni, A.S. and Nath, R.**, A comparison of radial dose functions for $^{103}$Pd, $^{125}$I, $^{145}$Sm, $^{241}$Am, $^{169}$Yb, $^{192}$Ir, and $^{137}$Cs brachytherapy sources, *Int. J. Radiat. Oncol. Biol. Phys.*, 22, 1125, 1992.

198. **Zellmer, D.L., Gillin, M.T., and Wilson, F.**, Microdosimetric single event spectra of ytterbium-169 compared with commonly used brachytherapy sources and teletherapy beams, *Int. J. Radiat. Oncol. Biol. Phys.*, 23, 627, 1992.

Chapter 5

# DATA ON GENERAL RADIATION PHYSICS AND DOSIMETRY

**LIST OF TABLES**

## TABLE 5.1
## The International System of Units (SI)

### Base SI Units

| Physical Quantity | Name | Symbol |
|---|---|---|
| Length | Meter | m |
| Mass | Kilogram | kg |
| Time | Second | s |
| Electric current | Ampere | A |
| Thermodynamic temperature | Kelvin | K |
| Amount of substance | Mole | mol |
| Luminous intensity | Candela | cd |

### Supplementary SI Units

| Physical Quantity | Name | Symbol |
|---|---|---|
| Plane angle | Radian | rad |
| Solid angle | Steradian | sr |

### Derived SI Units with Special Names

| Physical Quantity | Name | Symbol |
|---|---|---|
| Frequency | Hertz | Hz |
| Energy | Joule | J |
| Force | Newton | N |
| Power | Watt | W |
| Pressure | Pascal | Pa |
| Electric charge | Coulomb | C |
| Electric potential difference | Volt | V |
| Electric resistance | Ohm | $\Omega$ |
| Electric conductance | Siemens | S |
| Electric capacitance | Farad | F |
| Magnetic flux | Weber | Wb |
| Inductance | Henry | H |
| Magnetic flux density (magnetice induction) | Tesla | T |
| Luminous flux | Lumen | lm |
| Illumination | Lux | lx |

**TABLE 5.2**
**Prefixes Used with SI Units**

| Factor | Prefix | Symbol |
|---|---|---|
| $10$ | Deca- | da |
| $10^2$ | Hecto- | h |
| $10^3$ | Kilo- | k |
| $10^6$ | Mega- | M |
| $10^9$ | Giga- | G |
| $10^{12}$ | Tera- | T |
| $10^{15}$ | Peta- | P |
| $10^{18}$ | Exa- | E |
| $10^{-1}$ | Deci- | d |
| $10^{-2}$ | Centi- | c |
| $10^{-3}$ | Milli- | m |
| $10^{-6}$ | Micro- | $\mu$ |
| $10^{-9}$ | Nano- | n |
| $10^{-12}$ | Pico | p |
| $10^{-15}$ | Femto | f |
| $10^{-18}$ | Atto- | a |

**TABLE 5.3**
**The Greek Alphabet**

| Alpha | A | $\alpha$ |
|---|---|---|
| Beta | B | $\beta$ |
| Gamma | $\Gamma$ | $\gamma$ |
| Delta | $\Delta$ | $\delta$ |
| Epsilon | E | $\varepsilon$ |
| Zeta | Z | $\zeta$ |
| Eta | H | $\eta$ |
| Theta | $\Theta$ | $\theta$ |
| Iota | I | $\iota$ |
| Kappa | K | $\kappa$ |
| Lambda | $\Lambda$ | $\lambda$ |
| Mu | M | $\mu$ |
| Nu | N | $\nu$ |
| Xi | $\Xi$ | $\xi$ |
| Omicron | O | $o$ |
| Pi | $\Pi$ | $\pi$ |
| Rho | P | $\rho$ |
| Sigma | $\Sigma$ | $\sigma$ |
| Tau | T | $\tau$ |
| Upsilon | Y | $\upsilon$ |
| Phi | $\Phi$ | $\phi$ |
| Chi | X | $\chi$ |
| Psi | $\Psi$ | $\psi$ |
| Omega | $\Omega$ | $\omega$ |

## TABLE 5.4
## Other Units, Constants, and Conversion Factors

### Atomic

| | | |
|---|---|---|
| Mass of $^{12}C$ atom | = | Exactly 12 amu |
| 1 Atomic mass unit | = | $1.661 \times 10^{-27}$ kg |
| Proton charge | = | $4.803 \times 10^{-10}$ esu |
| Neutron rest mass ($M_n$) | = | 1.008665 amu |
| Mass of hydrogen atom ($M_h$) | = | 1.007825 amu |
| Avogadro constant ($N_o$) | = | $6.022 \times 10^{23}$/mole |

### Mass

| | | |
|---|---|---|
| 1 Kilogram (kg) | = | 2.205 pounds |
| 1 Pound (lb) | = | 453.592 g |
| 1 Metric ton (t) | = | 1000 kg |

### Energy

| | | |
|---|---|---|
| 1 Joule | = | $10^7$ ergs |
| 1 Electron volt (eV) | = | $1.602 \times 10^{-19}$ joules |
| | = | $1.07 \times 10^{-9}$ amu |
| 1 Watt | = | 1 joule/sec |
| 1 Horsepower | = | 746 watts |
| 1 British thermal unit (BTU) | = | 1055.18 joules |
| | = | 252 calories |
| 1 Joule = 0.24 calories | | |
| 1 Calorie | = | 4.18 joules |
| 1 Atomic mass unit | = | $9.31 \times 10^2$ MeV |

### Electronic

| | | |
|---|---|---|
| Electronic charge (mks system) | = | $1.6021 \times 10^{-19}$ coulomb |
| Electronic rest mass ($M_o$) | = | $9.1091 \times 10^{-28}$ g |
| | = | $5.486 \times 10^{-4}$ amu |
| Electronic charge to mass ratio (e/m) | = | $1.76 \times 10^{11}$ coul/kg |
| Electronic rest energy ($mc^2$) | = | 0.51101 MeV |
| 1 Coulomb | = | $6.28 \times 10^{18}$ electronic charge |

### Area

| | | |
|---|---|---|
| 1 Hectare (ha) | = | $10^4$ m$^2$ |
| 1 Barn (b) | = | $10^{-24}$ cm$^2$ |
| 1 Millibarn (mb) | = | $10^{-27}$ cm$^2$ |

### Other

| | | |
|---|---|---|
| Planck's constant (h) | = | $6.6256 \times 10^{-34}$ J-s |
| Speed of light in vacuum (c) | = | $2.9979 \times 10^{10}$ cm/sec |
| Photon wavelength at 1 eV | = | $12,295 \times 10^{-8}$ cm |
| Standard acceleration of gravity ($G_n$) | = | 9.80665 m/sec$^2$ |
| Standard atmospheric pressure (atm) | = | $1.01325 \times 10^5$ N/m$^2$ (or Pascal) |
| | = | 1.03323 kg/cm$^2$ |

## TABLE 5.5
## Radiation Quantities and Units

| Quantity | Symbol | SI Units | Special Unit | Conversion Factor |
|---|---|---|---|---|
| Exposure | X | $C\,kg^{-1}$ | Roentgen (R) | $1R = 2.58 \times 10^{-4}\,C\,kg^{-1}$ |
| Absorbed dose | D | $J\,kg^{-1}$ | Gray (Gy) | $1\,Gy = 1\,J\,kg^{-1}$ <br> $1\,Gy = 100\,rads$ |
| Decay constant | $\lambda$ | $s^{-1}$ | — | Half life $= (\ln^2)\,\lambda^{-1}$ |
| Activity | A | $s^{-1}$ | Becquerel (Bq) | $1\,Bq = 1\,s^{-1}$ <br> $1\,Ci = 3.7 \times 10^{10}\,Bq$ <br> $1\,mCi = 37\,MBq$ |
| Air kerma rate constant | $\Gamma_\delta$ | $m^2\,J\,kg^{-1}$ | $m^2\,Gy\,Bq^{-1}\,s^{-1}$ | $1\,m^2\,J\,kg^{-1} = 3.7 \times 10^{12}\,rad\,m^2\,Ci^{-1}\,s^{-1}$ <br> $1\,\mu Gy\,m^2\,GBq^{-1}\,hr^{-1} = 0.037\,cGy\,cm^2\,mCi^{-1}\,hr^{-1}$ |
| Dose equivalent | H | $J\,kg^{-1}$ | Sievert (Sv) | $1\,Sv = 1\,J\,kg^{-1}$ <br> $1\,Sv = 100\,rem$ |

**TABLE 5.6**
**Element Table**

| Atomic Number | Symbol | Name | Form At STP | Atomic* Weight | Density[†] (gm/cm³) | Melting Point °K | Boiling Point °K |
|---|---|---|---|---|---|---|---|
| 1 | H | Hydrogen | Gas | 1.008 | 0.0899 | 14.025 | 20.268 |
| 2 | He | Helium | Gas | 4.003 | 0.1787 | 0.95 | 4.215 |
| 3 | Li | Lithium | Solid | 6.941 | 0.53 | 453.7 | 1615 |
| 4 | Be | Beryllium | Solid | 9.012 | 1.85 | 1560 | 2745 |
| 5 | B | Boron | Solid | 10.810 | 2.34 | 2300 | 4275 |
| 6 | C | Carbon | Solid | 12.011 | 2.62 | 4100 | 4470 |
| 7 | N | Nitrogen | Gas | 14.007 | 1.251 | 63.14 | 77.35 |
| 8 | O | Oxygen | Gas | 15.999 | 1.43 | 50.35 | 90.18 |
| 9 | F | Fluorine | Gas | 18.998 | 1.696 | 53.48 | 84.95 |
| 10 | Ne | Neon | Gas | 20.179 | 0.901 | 24.553 | 27.096 |
| 11 | Na | Sodium | Solid | 22.990 | 0.97 | 371.0 | 1156 |
| 12 | Mg | Magnesium | Solid | 24.305 | 1.74 | 922 | 1363 |
| 13 | Al | Aluminum | Solid | 26.981 | 2.70 | 933.25 | 2793 |
| 14 | Si | Silicon | Solid | 28.086 | 2.33 | 1685 | 3540 |
| 15 | P | Phosphorus | Solid | 30.974 | 1.82 | 317.3 | 550 |
| 16 | S | Sulfur | Solid | 32.060 | 2.07 | 388.36 | 717.75 |
| 17 | Cl | Chlorine | Gas | 35.453 | 3.17 | 172.16 | 239.1 |
| 18 | Ar | Argon | Gas | 39.948 | 1.784 | 83.81 | 87.30 |

\* Based upon Carbon-12.      † For gaseous elements the density is in g/l at 273 K and 1 atm.

## TABLE 5.6 (continued)
### Element Table

| Atomic Number | Symbol | Name | Form At STP | Atomic* Weight | Density† (gm/cm³) | Melting Point °K | Boiling Point °K |
|---|---|---|---|---|---|---|---|
| 19 | K | Potassium | Solid | 39.098 | 0.86 | 336.35 | 1032 |
| 20 | Ca | Calcium | Solid | 40.080 | 1.55 | 1112 | 1757 |
| 21 | Sc | Scandium | Solid | 44.956 | 3.0 | 1812 | 3104 |
| 22 | Ti | Titanium | Solid | 47.900 | 4.50 | 1943 | 3562 |
| 23 | V | Vanadium | Solid | 50.942 | 5.8 | 2175 | 3682 |
| 24 | Cr | Chromium | Solid | 51.996 | 7.19 | 2130 | 2945 |
| 25 | Mn | Manganese | Solid | 54.938 | 7.43 | 1517 | 2335 |
| 26 | Fe | Iron | Solid | 55.847 | 7.86 | 1809 | 3135 |
| 27 | Co | Cobalt | Solid | 58.933 | 8.90 | 1768 | 3201 |
| 28 | N | Nickel | Solid | 58.7 | 8.9 | 1726 | 3187 |
| 29 | Cu | Copper | Solid | 63.546 | 8.96 | 1357.6 | 2836 |
| 30 | Zn | Zinc | Solid | 65.38 | 7.14 | 692.73 | 1180 |
| 31 | Ga | Gallium | Liquid | 69.72 | 5.91 | 302.90 | 2478 |
| 32 | Ge | Germanium | Solid | 72.59 | 5.32 | 1210.4 | 3107 |
| 33 | As | Arsenic | Solid | 74.922 | 5.72 | 1081 | 876 |
| 34 | Se | Selenium | Solid | 78.96 | 4.80 | 494 | 958 |
| 35 | Br | Bromine | Liquid | 79.904 | 3.12 | 265.90 | 332.25 |
| 36 | Kr | Krypton | Gas | 83.8 | 3.74 | 115.78 | 119.80 |

* Based upon Carbon-12.    † For gaseous elements the density is in g/l at 273 K and 1 atm.

## TABLE 5.6 (continued)
### Element Table

| Atomic Number | Symbol | Name | Form At STP | Atomic* Weight | Density† (gm/cm³) | Melting Point °K | Boiling Point °K |
|---|---|---|---|---|---|---|---|
| 37 | Rb | Rubidium | Solid | 85.468 | 1.53 | 312.64 | 961 |
| 38 | Sr | Strontium | Solid | 87.62 | 2.6 | 1041 | 1650 |
| 39 | Y | Yttrium | Solid | 88.906 | 4.5 | 1799 | 3611 |
| 40 | Zr | Zirconium | Solid | 91.22 | 6.49 | 2125 | 4682 |
| 41 | Nb | Niobium | Solid | 92.906 | 8.55 | 2740 | 5017 |
| 42 | Mo | Molybdenum | Solid | 95.94 | 10.2 | 2890 | 4912 |
| 43 | Tc | Technetium | — | 98.906 | 11.5 | 2473 | 4538 |
| 44 | Ru | Ruthenium | Solid | 101.07 | 12.2 | 2523 | 4423 |
| 45 | Rh | Rhodium | Solid | 102.905 | 12.4 | 2236 | 3970 |
| 46 | Pd | Palladium | Solid | 106.4 | 12.0 | 1825 | 3237 |
| 47 | Ag | Silver | Solid | 107.868 | 10.5 | 1234 | 2436 |
| 48 | Cd | Cadmium | Solid | 112.41 | 8.65 | 594.18 | 1040 |
| 49 | In | Indium | Solid | 114.82 | 7.31 | 429.76 | 2346 |
| 50 | Sn | Tin | Solid | 118.69 | 7.30 | 505.06 | 2876 |
| 51 | Sb | Antimony | Solid | 121.75 | 6.68 | 904 | 1860 |
| 52 | Te | Tellurium | Solid | 127.60 | 6.24 | 722.65 | 1261 |
| 53 | I | Iodine | Solid | 126.904 | 4.92 | 386.7 | 458.4 |
| 54 | Xe | Xenon | Gas | 131.30 | 5.89 | 161.36 | 165.03 |

* Based upon Carbon-12.     † For gaseous elements the density is in g/l at 273 K and 1 atm.

TABLE 5.6 (continued)
Element Table

| Atomic Number | Symbol | Name | Form At STP | Atomic* Weight | Density[†] (gm/cm³) | Melting Point °K | Boiling Point °K |
|---|---|---|---|---|---|---|---|
| 55 | Cs | Cesium | Liquid | 132.905 | 1.87 | 301.55 | 944 |
| 56 | Ba | Barium | Solid | 137.33 | 3.5 | 1002 | 2171 |
| 57 | La | Lanthanum | Solid | 138.905 | 6.7 | 1193 | 3730 |
| 58 | Ce | Cerium | Solid | 140.12 | 6.78 | 1071 | 3699 |
| 59 | Pr | Praseodymium | Solid | 140.908 | 6.77 | 1204 | 3785 |
| 60 | Nd | Neodymium | Solid | 144.24 | 7.00 | 1289 | 3341 |
| 61 | Pm | Promethium | — | 145 | 6.475 | 1204 | 3785 |
| 62 | Sm | Samarium | Solid | 150.4 | 7.54 | 1345 | 2064 |
| 63 | Eu | Europium | Solid | 151.96 | 5.26 | 1090 | 1870 |
| 64 | Gd | Gadolinium | Solid | 157.25 | 7.89 | 1585 | 3539 |
| 65 | Tb | Terbium | Solid | 158.925 | 8.27 | 1630 | 3496 |
| 66 | Dy | Dysprosium | Solid | 162.50 | 8.54 | 1682 | 2835 |
| 67 | Ho | Holmium | Solid | 164.930 | 8.80 | 1743 | 2968 |
| 68 | Er | Erbium | Solid | 167.26 | 9.05 | 1795 | 3136 |
| 69 | Tm | Thulium | Solid | 168.934 | 9.33 | 1818 | 2220 |
| 70 | Yb | Ytterbium | Solid | 173.04 | 6.98 | 1097 | 1467 |
| 71 | Lu | Lutetium | Solid | 174.967 | 9.84 | 1936 | 3668 |
| 72 | Hf | Hafnium | Solid | 178.49 | 13.1 | 2500 | 4876 |

* Based upon Carbon-12.   † For gaseous elements the density is in g/l at 273 K and 1 atm.

**TABLE 5.6 (continued)**
**Element Table**

| Atomic Number | Symbol | Name | Form At STP | Atomic* Weight | Density[†] (gm/cm$^3$) | Melting Point °K | Boiling Point °K |
|---|---|---|---|---|---|---|---|
| 73 | Ta | Tantalum | Solid | 180.947 | 16.6 | 3287 | 5731 |
| 74 | W | Tungsten | Solid | 183.85 | 19.3 | 3680 | 5828 |
| 75 | Re | Rhenium | Solid | 186.207 | 21.0 | 3453 | 5869 |
| 76 | Os | Osmium | Solid | 190.2 | 22.4 | 3300 | 5285 |
| 77 | Ir | Iridium | Solid | 192.22 | 22.5 | 2716 | 4701 |
| 78 | Pt | Platinum | Solid | 195.09 | 21.4 | 2045 | 4100 |
| 79 | Au | Gold | Solid | 196.966 | 19.3 | 1337.58 | 3130 |
| 80 | Hg | Mercury | Liquid | 200.59 | 13.53 | 234.28 | 630 |
| 81 | Tl | Thallium | Solid | 204.37 | 11.85 | 577 | 1746 |
| 82 | Pb | Lead | Solid | 207.2 | 11.4 | 600.6 | 2023 |
| 83 | Bi | Bismuth | Solid | 208.980 | 9.8 | 544.52 | 1837 |
| 84 | Po | Polonium | Solid | 209 | 9.4 | 527 | 1235 |
| 85 | At | Astatine | Solid | 210 | — | 575 | 610 |
| 86 | Rn | Radon | Gas | 222 | 9.91 | 202 | 211 |
| 87 | Fr | Francium | Liquid | 223 | — | 300 | 950 |

\* Based upon Carbon-12.      † For gaseous elements the density is in g/l at 273 K and 1 atm.

**TABLE 5.6 (continued)**
**Element Table**

| Atomic Number | Symbol | Name | Form At STP | Atomic* Weight | Density† (gm/cm³) | Melting Point °K | Boiling Point °K |
|---|---|---|---|---|---|---|---|
| 88 | Ra | Radium | Solid | 226.025 | 5.0 | 973 | 1809 |
| 89 | Ac | Actinium | Solid | 227.028 | 10.07 | 1323 | 3473 |
| 90 | Th | Thorium | Solid | 232.038 | 11.7 | 2028 | 5061 |
| 91 | Pa | Protactinium | Solid | 231.036 | 15.4 | — | — |
| 92 | U | Uranium | Solid | 238.029 | 18.90 | 1405 | 4407 |
| 93 | Np | Neptunium | — | 237.048 | 20.4 | 910 | — |
| 94 | Pu | Plutonium | — | 244 | 19.8 | 913 | 3503 |
| 95 | Am | Americium | — | 243 | 13.6 | 1268 | 2880 |
| 96 | Cm | Curium | — | 247 | 13.511 | 1340 | — |
| 97 | Bk | Berkelium | — | 247 | 14 (est) | — | — |
| 98 | Cf | Californium | — | 251 | — | 900 | — |
| 99 | Es | Einsteinium | — | 252 | — | — | — |
| 100 | Fm | Fermium | — | 257 | — | — | — |
| 101 | Md | Mendelevium | — | 258 | — | — | — |
| 102 | No | Nobelium | — | 259 | — | — | — |
| 103 | Lr | Lawrencium | — | 260 | — | — | — |
| 104 | Unq | Unnilquadium | — | 261 | — | — | — |
| 105 | Unp | Unnilpentium | — | 262 | — | — | — |
| 106 | Unh | Unnilhexium | — | 263 | — | — | — |

\* Based upon Carbon-12.    † For gaseous elements the density is in g/l at 273 K and 1 atm.

**TABLE 5.7**
**Physical Properties of Some Phantom Materials**

| | Water | Acrylic | Polystyrene | ICRU Muscle | Solid Water |
|---|---|---|---|---|---|
| Composition | $H_2O$ | $C_5H_8O_2$ | $C_8H_8$ | — | — |
| Density (g/cm$^3$) | 1.00 | 1.17* | 1.04* | 1.00 | 1.015 |
| Ave Atomic No. (Z)$^\dagger$ | 7.22 | 6.24 | 5.62 | 7.10 | — |
| Ave Z relative to water | 1.000 | 0.86 | 0.78 | 0.98 | — |
| Electron density (e$^-$/g) | $3.346\times10^{23}$ | $3.253\times10^{23}$ | $3.243\times10^{23}$ | $3.32\times10^{23}$ | — |
| Electron concentration (e$^-$/cm$^3$) | $3.346\times10^{23}$ | $3.806\times10^{23}$ | $3.373\times10^{23}$ | $3.32\times10^{23}$ | — |

\* Nominal values
$^\dagger$ The average atomic number is calculated by weighing the component atomic numbers by parts by weight.

**TABLE 5.8**
**Physical Properties of Adult Human Body Tissues**

| Tissue | Elemental Composition (Percentage by Mass) | | | | | Mass Density (gm/cm$^3$) | Electron Density e$^-$/cm$^3$×10$^{23}$ |
|---|---|---|---|---|---|---|---|
| | H | C | N | O | Others* | | |
| Adipose tissue | 11.4 | 59.8 | 0.7 | 27.8 | — | 0.950 | 3.18 |
| Blood (whole) | 10.2 | 11.0 | 3.3 | 74.5 | — | 1.060 | 3.51 |
| Brain | 10.7 | 14.5 | 2.2 | 71.2 | — | 1.04 | 3.46 |
| Breast (mammary gland) | 10.6 | 33.2 | 3.0 | 52.7 | — | 1.020 | 3.39 |
| Eye lens | 9.6 | 19.5 | 5.7 | 64.6 | — | 1.070 | 3.53 |
| GI tract (intestine) | 10.6 | 11.5 | 2.2 | 75.1 | — | 1.030 | 3.42 |
| Heart (blood filled) | 10.3 | 12.1 | 3.2 | 73.4 | — | 1.060 | 3.51 |
| Kidney | 10.3 | 13.2 | 3.0 | 72.4 | — | 1.050 | 3.48 |
| Liver | 10.2 | 13.9 | 3.0 | 71.6 | — | 1.060 | 3.51 |
| Lung | 10.3 | 10.5 | 3.1 | 74.9 | — | 1.050 -deflated | 3.48 |
| | | | | | | 0.260 -inflated | 0.862 |
| Muscle(skeletal) | 10.2 | 14.3 | 3.4 | 71.0 | — | 1.050 | 3.48 |
| Ovary | 10.5 | 9.3 | 2.4 | 76.8 | — | 1.050 | 3.49 |
| Pancreas | 10.6 | 16.9 | 2.2 | 69.4 | — | 1.040 | 3.46 |
| Skeleton-cartilage | 9.6 | 9.9 | 2.2 | 74.4 | 0.9 S | 1.100 | 3.62 |
| Skeleton-cortical bone | 3.4 | 15.5 | 4.2 | 43.5 | 10.3 P, 22.5 Ca | 1.920 | 5.95 |
| Skeleton-red marrow | 10.5 | 41.4 | 3.4 | 43.9 | — | 1.030 | 3.42 |
| Skeleton-yellow marrow | 11.5 | 64.4 | 0.7 | 23.1 | — | 0.980 | 3.28 |
| Skin | 10.0 | 20.4 | 4.2 | 64.5 | — | 1.090 | 3.60 |
| Spleen | 10.3 | 11.3 | 3.2 | 74.1 | — | 1.060 | 3.51 |
| Testis | 10.6 | 9.9 | 2.0 | 76.6 | — | 1.040 | 3.46 |
| Thyroid | 10.4 | 11.9 | 2.4 | 74.5 | — | 1.050 | 3.48 |

* Very small amounts of Na, P, S, Cl, K, Mg, etc. found in tissues are not listed here. See Ref. 8 for further details.

From ICRU Report 44, International Commission on Radiation Units and Measurement, Bethesda, MD, 1989. With permission.

Handbook of Dosimetry Data for Radiotherapy

## TABLE 5.9
## Effective Atomic Numbers of Human Body Tissues (Error: ±0.2)

| | | 1 MeV | 5 MeV | 10 MeV | 20 MeV | 50 MeV |
|---|---|---|---|---|---|---|
| Bone | Photons | 6.0 | 6.3 | 6.7 | 7.1 | 7.5 |
| | Electrons | 6.0 | 6.3 | 6.4 | 6.6 | 6.7 |
| | He ions | 5.1 | 5.4 | 5.5 | 5.5 | 5.5 |
| Muscle | Photons | 3.4 | 3.6 | 3.8 | 4.0 | 4.4 |
| | Electrons | 3.5 | 3.7 | 3.9 | 4.1 | 4.3 |
| | He ions | 3.3 | 3.4 | 3.2 | 3.2 | 3.3 |
| Liver | Photons | 3.4 | 3.6 | 3.8 | 4.0 | 4.3 |
| | Electrons | 3.5 | 3.7 | 3.9 | 4.1 | 4.3 |
| | He ions | 3.3 | 3.4 | 3.1 | 3.2 | 3.3 |
| Spleen | Photons | 3.4 | 3.6 | 3.8 | 4.0 | 4.4 |
| | Electrons | 3.5 | 3.7 | 3.9 | 4.1 | 4.4 |
| | He ions | 3.3 | 3.4 | 3.2 | 3.2 | 3.3 |
| Mucin | Photons | 4.6 | 4.7 | 4.9 | 5.1 | 5.3 |
| | Electrons | 4.6 | 5.0 | 5.1 | 5.2 | 5.5 |
| | He ions | 4.5 | 4.6 | 4.4 | 4.4 | 4.4 |
| Water | Photons | 3.3 | 3.5 | 3.7 | 4.0 | 4.3 |
| | Electrons | 3.4 | 3.7 | 3.9 | 4.1 | 4.3 |
| | He ions | 3.2 | 3.3 | 3.1 | 3.2 | 3.2 |

From Parthasaradhi, K., Rao, B.M., and Prasad, S.G., Effective atomic numbers of biological materials in the energy region 1 to 50 MeV for photons, electrons, and He ions, *Med. Phys.*, 16, 653, 1989. With permission.

## TABLE 5.10
### Temperature-Pressure Correction Factors for Open-to-Air Ionization Chambers

| Pressure mm Hg | Temperature (degrees Celsius) | | | | | | | | | | | | |
|---|---|---|---|---|---|---|---|---|---|---|---|---|
| | 19.0 | 19.5 | 20.0 | 20.5 | 21.0 | 21.5 | 22.0 | 22.5 | 23.0 | 23.5 | 24.0 | 24.5 | 25.0 |
| 710 | 1.060 | 1.061 | 1.063 | 1.065 | 1.067 | 1.069 | 1.070 | 1.072 | 1.074 | 1.076 | 1.078 | 1.079 | 1.081 |
| 712 | 1.057 | 1.058 | 1.060 | 1.062 | 1.064 | 1.066 | 1.067 | 1.069 | 1.071 | 1.073 | 1.075 | 1.076 | 1.078 |
| 714 | 1.054 | 1.055 | 1.057 | 1.059 | 1.061 | 1.063 | 1.064 | 1.066 | 1.068 | 1.070 | 1.072 | 1.073 | 1.075 |
| 716 | 1.051 | 1.052 | 1.054 | 1.056 | 1.058 | 1.060 | 1.061 | 1.063 | 1.065 | 1.067 | 1.069 | 1.070 | 1.072 |
| 718 | 1.048 | 1.050 | 1.051 | 1.053 | 1.055 | 1.057 | 1.058 | 1.060 | 1.062 | 1.064 | 1.066 | 1.067 | 1.069 |
| 720 | 1.045 | 1.047 | 1.048 | 1.050 | 1.052 | 1.054 | 1.056 | 1.057 | 1.059 | 1.061 | 1.063 | 1.064 | 1.066 |
| 722 | 1.042 | 1.044 | 1.045 | 1.047 | 1.049 | 1.051 | 1.053 | 1.054 | 1.056 | 1.058 | 1.060 | 1.062 | 1.063 |
| 724 | 1.039 | 1.041 | 1.043 | 1.044 | 1.046 | 1.048 | 1.050 | 1.052 | 1.053 | 1.055 | 1.057 | 1.059 | 1.060 |
| 726 | 1.036 | 1.038 | 1.040 | 1.042 | 1.043 | 1.045 | 1.047 | 1.049 | 1.050 | 1.052 | 1.054 | 1.056 | 1.057 |
| 728 | 1.033 | 1.035 | 1.037 | 1.039 | 1.040 | 1.042 | 1.044 | 1.046 | 1.047 | 1.049 | 1.051 | 1.053 | 1.055 |
| 730 | 1.031 | 1.032 | 1.034 | 1.036 | 1.038 | 1.039 | 1.041 | 1.043 | 1.045 | 1.046 | 1.048 | 1.050 | 1.052 |
| 732 | 1.028 | 1.029 | 1.031 | 1.033 | 1.035 | 1.036 | 1.038 | 1.040 | 1.042 | 1.044 | 1.045 | 1.047 | 1.049 |
| 734 | 1.025 | 1.027 | 1.028 | 1.030 | 1.032 | 1.034 | 1.035 | 1.037 | 1.039 | 1.041 | 1.042 | 1.044 | 1.046 |
| 736 | 1.022 | 1.024 | 1.026 | 1.027 | 1.029 | 1.031 | 1.033 | 1.034 | 1.036 | 1.038 | 1.040 | 1.041 | 1.043 |
| 738 | 1.019 | 1.021 | 1.023 | 1.025 | 1.026 | 1.028 | 1.030 | 1.032 | 1.033 | 1.035 | 1.037 | 1.039 | 1.040 |
| 740 | 1.017 | 1.018 | 1.020 | 1.022 | 1.024 | 1.025 | 1.027 | 1.029 | 1.031 | 1.032 | 1.034 | 1.036 | 1.037 |
| 742 | 1.014 | 1.016 | 1.017 | 1.019 | 1.021 | 1.023 | 1.024 | 1.026 | 1.028 | 1.029 | 1.031 | 1.033 | 1.035 |
| 744 | 1.011 | 1.013 | 1.015 | 1.016 | 1.018 | 1.020 | 1.022 | 1.023 | 1.025 | 1.027 | 1.028 | 1.030 | 1.032 |
| 746 | 1.008 | 1.010 | 1.012 | 1.014 | 1.015 | 1.017 | 1.019 | 1.020 | 1.022 | 1.024 | 1.026 | 1.027 | 1.029 |
| 748 | 1.006 | 1.007 | 1.009 | 1.011 | 1.013 | 1.014 | 1.016 | 1.018 | 1.019 | 1.021 | 1.023 | 1.025 | 1.026 |
| 750 | 1.003 | 1.005 | 1.006 | 1.008 | 1.010 | 1.012 | 1.013 | 1.015 | 1.017 | 1.018 | 1.020 | 1.022 | 1.024 |
| 752 | 1.000 | 1.002 | 1.004 | 1.006 | 1.007 | 1.009 | 1.011 | 1.012 | 1.014 | 1.016 | 1.017 | 1.019 | 1.021 |
| 754 | 0.998 | 0.999 | 1.001 | 1.003 | 1.005 | 1.006 | 1.008 | 1.010 | 1.011 | 1.013 | 1.015 | 1.016 | 1.018 |
| 756 | 0.995 | 0.997 | 0.998 | 1.000 | 1.002 | 1.004 | 1.005 | 1.007 | 1.009 | 1.010 | 1.012 | 1.014 | 1.016 |
| 758 | 0.992 | 0.994 | 0.996 | 0.998 | 0.999 | 1.001 | 1.003 | 1.004 | 1.006 | 1.008 | 1.009 | 1.011 | 1.013 |
| 760 | 0.990 | 0.992 | 0.993 | 0.995 | 0.997 | 0.998 | 1.000 | 1.002 | 1.003 | 1.005 | 1.007 | 1.008 | 1.010 |

Correction Factor = $(760/P) * \{(273.15+T)/295.15\}$

# REFERENCES

1. **Cohen, E.R. and Taylor, B.N.,** The fundamental physical constants, *Phys. Today,* 45, 9, 1992.
2. **Constantinou, C., Attix, F.H., and Paliwal, B.R.,** A solid water phantom material for radiotherapy x-ray and γ-ray beam calibrations, *Med. Phys.,* 9, 436, 1982
3. **Parthasaradhi, K., Rao, B.M., and Prasad, S.G.,** Effective atomic numbers of biological materials in the energy region 1 to 50 MeV for photons, electrons, and He ions, *Med. Phys.,* 16, 653, 1989.
4. *Photon, Electron, Proton and Neutron Interaction Data for Body Tissues,* ICRU Report 46, International Commission on Radiation Units and Measurements, Bethesda, MD, 1992.
5. *Radiation Quantities and Units,* ICRU Report 33, International Commission on Radiation Units and Measurements, Bethesda, MD, 1980.
6. **Reft, C.S.,** Output calibration in solid water for high energy photon beams, *Med. Phys.,* 16, 299, 1989.
7. Sargent-Welch Scientific Co., Skokie, IL, personal communication, 1992.
8. *Tissue Substitutes in Radiation Dosimetry and Measurement,* ICRU Report 44, International Commission on Radiation Units and Measurements, Bethesda, MD, 1989.
9. **van Assendelft, O.W.,** The international system of units (SI) in historical perspective, *Am. J. Public Health,* 77, 1400, 1987.

# INDEX